BIOSECUR

Biosecurity is the assessment and management of potentially dangerous infectious diseases, quarantined pests, invasive (alien) species, living modified organisms and biological weapons. It is a holistic concept of direct relevance to the sustainability of agriculture, food safety and the protection of human populations (including bioterrorism), the environment and biodiversity. Biosecurity is a relatively new concept that has become increasingly prevalent in academic, policy and media circles, and needs a more comprehensive and inter-disciplinary approach to take into account mobility, globalization and climate change.

In this introductory volume, biosecurity is presented as a governance approach to a set of concerns that span the protection of indigenous biological organisms, agricultural systems and human health from invasive pests and diseases. It describes the ways in which biosecurity is understood and theorized in different subject disciplines, including anthropology, political theory, ecology, geography and environmental management. It examines the different scientific and knowledge practices connected to biosecurity governance, including legal regimes, ecology, risk management and alternative knowledges. The geopolitics of biosecurity is considered in terms of health, biopolitics and trade governance at the global scale. Finally, biosecurity as an approach to actively secure the future is assessed in the context of future risk and uncertainties, such as globalization and climate change.

Andrew Dobson is Professor of Politics at Keele University, UK.

Kezia Barker is Lecturer in Science and Environmental Studies at Birkbeck, University of London, UK.

Sarah L. Taylor is Lecturer in Ecology and Programme Director of Biology at Keele University, UK.

BIOSECURITY

The socio-politics of invasive
species and infectious diseases

Edited by
Andrew Dobson,
Kezia Barker and Sarah L. Taylor

Routledge
Taylor & Francis Group
LONDON AND NEW YORK

First published 2013
by Routledge
2 Park Square, Milton Park, Abingdon, Oxon OX14 4RN

Simultaneously published in the USA and Canada
by Routledge
711 Third Avenue, New York, NY 10017

Routledge is an imprint of the Taylor & Francis Group, an informa business

British Library Cataloguing-in-Publication Data
A catalogue record for this book is available from the British Library

Library of Congress Cataloging-in-Publication Data
Biosecurity : the socio-politics of invasive species and infectious diseases / edited by Andrew Dobson,
Kezia Barker, Sarah Taylor.
pages cm
Includes bibliographical references and index.
1. Biosecurity- -Political aspects. 2. Nonindigenous pests- -Control- -Political aspects.
3. Communicable diseases- -Prevention- -Political aspects. I. Dobson, Andrew. II. Barker, Kezia.
III. Taylor, Sarah L.
JZ5865.B56B576 2013
333.95'23- -dc23
2013001294

ISBN: 978-0-415-53476-5 (hbk)
ISBN: 978-0-415-53477-2 (pbk)
ISBN: 978-0-203-11311-0 (ebk)

Typeset in Goudy
by Taylor & Francis Books

MIX
Paper from
responsible sources
FSC   FSC® C013056
www.fsc.org

Printed and bound in Great Britain by
TJ International Ltd, Padstow, Cornwall

Sarah dedicates this book to her nana, Mary Taylor (1917–2010).

Kezia dedicates it to her grandparents, Sallie and Roy Barker, as one more great-grandchild inherits their lullabies.

Andy dedicates it to the memory of his mother, Jean Dobson (née Kirkbride) (1922–2006).

CONTENTS

CONTRIBUTORS

Kezia Barker is Lecturer in Science and Environmental Studies at Birkbeck, University of London, UK, where she convenes the undergraduate module 'Environment and Security'. She is a geographer with research interests in the intersection between environmental governance and political and cultural associations to nature. Kezia has specialized in understanding how natures and human–nonhuman associations are negotiated by the complex machineries of environmental regulation, and how that apparatus is challenged and responds to mobility, uncertainty and change. Her ongoing research focuses on biosecurity policy-making, surveillance and enforcement practices in New Zealand, the UK and Galápagos. Kezia is the editor of a special issue 'Infectious Insecurities' in the journal *Health and Place*, a critical exploration of the 2009 H1N1 epidemic. Other recent publications include a discussion of the reconfiguration of citizenship through biosecurity policy in the journal *Transactions of the Institute of British Geographers*. Kezia is the co-investigator of the ESRC seminar series 'The Socio-Politics of Biosecurity: Science, Policy and Practice', and co-editor of this book. She lives in Manchester with her husband and their young son.

Bruce Braun is Professor of Geography at the University of Minnesota, USA. He is the author of *The Intemperate Rainforest: Nature, Culture and Power on Canada's West Coast* (University of Minnesota Press, 2002) and co-editor of *Political Matter: Technoscience, Democracy and Public Life* (with Sarah Whatmore, University of Minnesota Press, 2010), among other edited books and articles. His recent work includes essays on the biopolitics of biosecurity, new materialisms and the politics of resilient cities. He is currently an editor of the *Annals of the Association of American Geographers* and co-editor of *Resilience: International Policies, Practices and Discourses*.

Henry Buller is Professor of Geography at Exeter University, UK, having formerly worked at the universities of Paris X and Paris VII, France. He has written widely in books, papers and reports on human/animal relations, animal welfare, environmental and nature management and rural development. He has recently been Principal Investigator on two major ESRC-funded research projects – 'Eating Biodiversity' (2005–2007) and 'Understanding Human Behaviour through Human/Animal Interaction' (2009–2010) as well as being Co-investigator on the EU-funded 'Welfare Quality' programme (2005–2010). He served as an appointed member of the Comité d'Orientation, de Recherche et de Prospective of the French Parcs Naturels Régionaux (2007–2011) and is an appointed member of the UK's Farm Animal Welfare Committee.

Andrew Dobson is Professor of Politics at Keele University, UK. He is the author of *Green Political Thought* (4th edition, 2006) and of *Citizenship and the Environment* (Oxford University Press, 2003), among other edited books, monographs and papers. He is Principal Investigator on a 2011–2013 ESRC/EPSRC-funded project on 'Reducing Energy Consumption Through Community Knowledge Networks' (RECCKN) (www.esci.keele.ac.uk/recckn). He is also a Leverhulme Research Fellow, writing a book provisionally entitled *Listening for Democracy*, which is due for publication by Oxford University Press in 2014. He co-wrote the England and Wales Green Party Manifesto in 2010 and is a founder member of the thinktank Green House (www.greenhousethinktank.org). His website is www.andrewdobson.com.

Andrew Donaldson has a background in natural and social sciences. Since 2001 a major strand of his research has been the management of animal disease risk, on which he has published extensively, including key early papers on the geography and politics of biosecurity. In 2008 he acted as an expert advisor on animal health policy for the UK's National Audit Office. His work has also introduced the consideration of nonhuman life into the interdisciplinary field of surveillance studies. Andrew is currently a Senior Lecturer in Newcastle University's School of Architecture, Planning and Landscape and continues to research the everyday management of natural hazards and biological infrastructure.

Gareth Enticott is a Senior Lecturer in Rural Geography at Cardiff University, UK. His research focuses on the social impacts of animal disease, specifically in relation to bovine tuberculosis. He has led ESRC- and Defra-funded research projects examining how farmers' understandings of animal disease shape their biosecurity practices. His most recent work examines how animal disease and veterinary expertise are shaped by organizational cultures, regulatory structures and governmental regimes in the UK and New Zealand.

Juliet J. Fall is an Anglo-Swiss political and environmental geographer working on questions of biosecurity, borders and national identity. She is interested in how discourses of nature are enrolled into politics and explores how apparently progressive references to the environment are subverted to underpin reactionary politics. Recent research has dealt with invasive species, 'natural borders', and urban gardening. Author of *Drawing the Line: Nature, Hybridity and Politics in Transboundary Spaces* (Ashgate, 2005); and co-editor of *Aux frontières de l'Animal: mises en scènes et reflexivités* (Droz, 2012) with A. Dubied and D. Gerber, she is a Professor at the University of Geneva, in Switzerland. She also works on the spaces and politics of knowledge production in a science studies perspective, and on the history of geography in anglophone and francophone contexts. She lives in Geneva with her family, including two small children, a sewing machine and guinea pigs.

Steve Hinchliffe is Professor of Human Geography at the University of Exeter, UK. He is the author of *Geographies of Nature* (Sage, 2008) and has edited numerous books on environment and society. He is the Principal Investigator on a UK Economic and Social Research Council-funded project running from 2009 to 2013 entitled 'Biosecurity Borderlands' (http://biosecurity-borderlands. org). He is also funded by the UK's Department for Environment, Food and Rural Affairs (Defra) to investigate the social aspects of bovine tuberculosis (2012–2014). Steve leads the University of Exeter's research strategy on Science, Technology and Culture and is an appointed member of the UK's Food Standards Agency (FSA) Social Science Research Committee. His website is http://geography.exeter.ac.uk/staff/index.php?web_id=Steve_Hinchliffe.

Alan Ingram is Senior Lecturer in the Department of Geography at University College London, UK, where he teaches political geography, geopolitics and security. He is co-editor of *Spaces of Security and Insecurity: Geographies of the War on Terror* (Ashgate, 2009) and in 2011–2012 held a British Academy Mid-Career Fellowship for the project 'Art, and War: Responses to Iraq'. From 2002 to 2004 he managed a policy research and development programme on health, foreign policy and security at the Nuffield Trust, and his academic research in this area focuses on intersections between governmentality, political economy and security, with particular reference to international responses to HIV/AIDS, global health security and global health diplomacy.

Filippa Lentzos is a Senior Research Fellow in the Department of Social Science, Health and Medicine at King's College London, UK. Originally trained in human sciences before switching to sociology, she is particularly interested in social, political and security aspects of advances in the life sciences. Her current research on the politics of bioterrorism is funded by an ESRC Mid-Career Fellowship.

John Mumford is Professor of Natural Resource Management at Imperial College London, UK. He has been an author on various reviews of UK, EU and international plant biosecurity related to risk assessment, risk management and standard-setting. In recent years he has been a leader in research projects with the EC, EFSA and UK Defra on developing processes to improve national and regional pest risk assessments for both agricultural pests and new organisms in the natural environment. This has also included horizon scanning/foresight studies to determine future threats from exotic organisms. He works in a WTO/STDF project to improve competence and confidence on phytosanitary issues in trade negotiations. He chairs the Great Britain Non-native Species Risk Analysis Panel. His website is www.imperial.ac.uk/people/j.mumford.

Opi Outhwaite is Senior Lecturer in Law at the University of Greenwich, UK. Her research focuses on socio-legal, legislative and regulatory issues in biosecurity. Opi has authored a number of publications in this area and recently completed funded research on 'Legal Issues in Honey Bee Health and Conservation'. She is currently (2012–2013) Social Research Fellow in Animal Health at the Department of Environment, Food and Rural Affairs (Defra).

Clive Potter is Reader in Environmental Policy at Imperial College London, UK. A rural geographer by training, he has written widely about the political economy of agricultural policy and countryside change and has contributed to scholarly debate about the contemporary neoliberalization of rural nature. Clive's work in the field of biosecurity has examined the conflict between a growing international trade in plants and the desire to prevent tree disease epidemics which threaten tree health.

Brian Rappert is a Professor of Science, Technology and Public Affairs in the Department of Sociology and Philosophy at the University of Exeter, UK. His long-term interest has been the examination of the strategic management of information, particularly in relation to armed conflict. His books include *Controlling the Weapons of War: Politics, Persuasion, and the Prohibition of Inhumanity, Technology and Security* (editor); *Biotechnology, Security and the Search for Limits*; and *Education and Ethics in the Life Sciences* (ANU E-Press, 2012). More recently he has been interested in the social, ethical and epistemological issues associated with researching and writing about secrets, as in his book *Experimental Secrets* (Pluto Press, 2009) and *How to Look Good in a War* (Pluto Press, 2012).

Daniel Simberloff is the Nancy Gore Hunger Professor of Environmental Studies at the University of Tennessee, USA. His writings centre on ecology, biogeography, evolution and conservation biology; much of his research focuses on causes and consequences of biological invasions. His research projects are on insects, plants, fungi, birds and mammals. At the University of Tennessee he directs the Institute for Biological Invasions. He is Editor-in-Chief of *Biological Invasions*, Senior Editor of the *Encyclopedia of Biological Invasions* and has just completed a book, *What Everyone Needs to Know About Biological Invasions*, which is due for publication by Oxford University Press in 2013. He served on the United States National Science Board from 2000 to 2006 and is a member of the American Academy of Arts and Sciences and the US National Academy of Sciences. His website is eeb.bio.utk.edu/Simberloff.asp.

Sarah L. Taylor is Lecturer in Ecology and Programme Director of Biology at Keele University, UK. She is co-author on the Forestry Commission report on rhododendron invasion and control in woodland areas in Argyll and Bute, has written numerous peer-reviewed papers and is a regular book reviewer for the *British Ecological Society Bulletin*. This is her first book. She is Principal Investigator on an NERC-, ARSF- and FSF-funded project on 'Assessing the Impact of Environmental Factors on Invasive Potential of *Rhododendron* *ponticum* in Western Britain: Implications for Climate Change'. She is also an honorary research associate of the University of New Brunswick, USA; she serves on the Victoria Angling Club management committee and is education officer for GeoConservation Staffordshire. Her website is www.keele.ac.uk/lifesci/people/staylor.

Katy Wilkinson is Research Impact Officer at the University of Warwick, UK. Prior to this she held an academic fellowship in the Department for Environment, Food and Rural Affairs (Defra) and was a postdoctoral researcher on the UK Research Councils' Rural Economy and Land Use programme. She has published research on evidence-based policy-making, particularly in the areas of animal disease outbreaks; novel technologies in agriculture and human–livestock interactions; and methodological contributions on interdisciplinarity and interpretive policy analysis.

ACKNOWLEDGEMENTS

This book would not have been possible without the contributions of many people and organizations. First, we would like to thank the Economic and Social Research Council (ESRC) for funding the 2009–2011 seminar series on 'The Socio-Politics of Biosecurity: Science, Policy and Practice' (RES-451-26-0740) from which this book emerged. Second, thanks are due to the Research Institute for Social Sciences at Keele University and to the School of Social Sciences, History and Philosophy and the Birkbeck Institute of Social Research, Birkbeck College, University of London for logistical support in running the seminar series. In particular we would like to thank Helen Farrell and Tracey Wood at Keele.

We are very grateful to Earthscan, and especially to Tim Hardwick, for approaching us with the idea of editing a book drawing on the seminar series. Ashley Irons was helpful – and patient – as our conversations about the book cover unfolded. The book itself, of course, is all about its contributors. They have been unfailingly generous with their time and have responded with alacrity to each and every request from the editors. We have been fortunate indeed to work with such a committed group of scholars, especially in their determination to make their contributions accessible to an interdisciplinary audience.

Sarah and Kezia would like to thank Andrew Dobson for having the vision to bring together two early-career academics from contrasting disciplines to tackle the interdisciplinary subject of biosecurity. Andy owes a debt of gratitude to Kezia and Sarah for doing most of the intellectual and practical heavy lifting in regard to the seminar series and this book. Without their talent and enthusiasm neither would have been possible.

Finally we would like to thank Monique Martens for creating and maintaining a marvellous website, which is still available for consultation and which contains details of all the seminars and conversations on which this book draws (www.bbk. ac.uk/environment/biosecurity).

Keele and London
November 2012

Part I

FRAMING BIOSECURITY

1

INTRODUCTION
Interrogating bio-insecurities

Kezia Barker, Sarah L. Taylor and Andrew Dobson

Introduction to biosecurity: defining biosecurity threats

A leaflet drops through your letter box advising on ways to avoid contracting and spreading swine flu. Your bag is searched as you enter a country on the start of your holiday – and there's a fine of $200 for an undeclared apple. You drive across a disinfectant mat on a visit to a local farm. You call the council for advice on the Japanese knotweed (*Fallopia japonica*) spreading from your neighbour's garden. You switch to eating organic free-range chicken after reading about the role of industrial farming in the production of avian flu risk. You shudder as you remember the smell of burning carcasses in the English countryside back in 2001. In many different ways you may have encountered events, practices, procedures, narratives and knowledges contained within the complex issue of 'biosecurity'.

Just as Hajer (1995) described acid rain as an 'emblematic environmental issue' of the 1990s – an issue that functions as a metaphor for environmental problems at particular times – this collection demonstrates that biosecurity and associated issues, such as bioinvasion and nativism, are emblematic issues for the twenty-first century. This is because the study of biosecurity prompts reflection on a series of issues of great importance in contemporary society: the extension of security, selective territorializations against ever-increasing mobility, questions of localism and cosmopolitanism, the interaction of public and private domains in environmental management, and concerns over the construction of risk and the management of uncertain futures. Biosecurity provides a lens for interrogating these issues, and our response requires engagement from a host of disciplinary perspectives. This is what we aim to achieve in this book.

Across the social sciences, the burgeoning field of biosecurity studies, to which this edited collection is testament, is informed and driven by a number of theoretical currents. These include interests in governmentality and biopolitics (Collier et al., 2004; Cooper, 2006; Braun, 2007; Collier and Lakoff, 2008; Dillon and Lobo-Guerrero, 2008); questions of risk, uncertainty and indeterminacy (Donaldson, 2008; Hinchliffe, 2001; Fish et al., 2011); attention to nonhumans and co-produced networks, materiality, circulation and mobility (Clark, 2002; Ali and Keil, 2008; Braun, 2008; Wallace, 2009; Barker, 2010; Barker, in prep.); the interrogation of spatial processes of categorization and boundary-making (Donaldson and Wood, 2004; Barker, 2008; Tomlinson and Potter, 2010; Mather and Marshall, 2011); and geopolitical concerns with the interaction between nation states, processes of globalization, post-colonialism and modes of inequality (Farmer, 1999; King, 2002, 2003; Ingram, 2005, 2009; French, 2009; Sparke, 2009).

In some ways, crucial theoretical developments and new areas of enquiry have proceeded in a disease-driven way. HIV-AIDS, for example, has arguably marked the literature through attention to geopolitical concerns over the place of health in global governance as well as through questions of inequality (Elbe, 2005; Ingram, 2010). Scholarship emerging from reflections on the SARS epidemic drew attention to the globalized co-produced networks of disease exchange (Fidler, 2004; Ali and Keil, 2008; Braun, 2008). Meanwhile highly pathogenic avian flu has, through its construction as 'the next big thing', highlighted the future temporalities of disease governance, or what Samimian-Damash (2009) refers to as the 'pre-event configuration', the constellation of anticipatory discourses and practices through which biosecurity is enacted (Bingham and Hinchliffe, 2008). These diverse tendencies together demonstrate that no single theoretical lens is sufficient to fully encapsulate and respond to the critical issues raised by biosecurity, and reveal the capacity of biosecurity to act as fertile empirical ground for theoretical experimentation.

This combination of theoretical approaches, and the wide range of problem issues that biosecurity touches upon, makes it one of the most exciting and innovative areas of contemporary social research. The term itself, as Donaldson indicates, presents a 'semantic banquet' for geographers. 'The evocative "bio" prefix brings to mind the "relational ontologies" and "hybrid politics" … encapsulated in performances of nature, society and space … whilst the "security" element resonates powerfully with contemporary geopolitical concerns' (Donaldson, 2008, p. 1553).

For natural scientists researching biosecurity-related issues, the complexity resulting from interactions of physical, biological and human systems, and the ramifications of uncertain events such as climate change make this area of research challenging and enthralling. Increased computing power and the capacity of computer programmes such as Geographic Information Systems (GIS) to digitally represent disease or invasive species distributions, coupled with increased availability of remotely sensed data, have enabled scientists to analyse biosecurity issues on temporal and spatial scales that were previously impossible. Furthermore, predictions of species movements are increasing in accuracy as we get a better handle on bioclimatic-modelling (climate matching species' needs; see Araújo and Peterson, 2012 for a review) and an improved understanding of the physiological and behavioural changes of target organisms in a changing environment (e.g. Huey et al., 2012). Amidst all this, the potential for scientific advances to be misused is a cause for concern (The Royal Society and Wellcome Trust, 2004).

The National Science Advisory Board for Biosecurity, a panel of the US Department of Health and Human Services, makes recommendations on how to prevent biotechnology research from aiding terrorism without slowing scientific progress.

What biosecurity actually entails is itself up for debate, shifting across spatial, temporal and discursive contexts. Biosecurity can, in general terms, be described as the attempted management or control of unruly biological matter, ranging from microbes and viruses to invasive plants and animals. However, when delving deeper into the meanings and usages of biosecurity, it is immediately clear that variation exists in the term. The International Union for the Conservation of Nature (IUCN) provides an all-encompassing definition that places biosecurity within the domain of risk, describing a biosecurity threat as 'matters or activities which, individually or collectively, may constitute a biological risk to the ecological welfare or to the well-being of humans, animals or plants of a country' (IUCN 2000, p. 3). The term 'biosecurity' was largely unheard of in the UK prior to the 2001 foot-and-mouth disease (FMD) outbreak, during which it evolved from a reference to practices, such as cleansing and disinfecting, to the surveillance control of movement and spaces to stop the transmission of animal diseases within farming (Donaldson and Wood, 2004; see also Hinchliffe, 2001; Law, 2006). The latter led to enormous disruptions in both trade and tourism in affected areas (Irvine and Anderson, 2005). Internationally, in the post 9/11 era it has also come to be associated with the prevention of bioterrorism, laboratory biosafety and the spread of apocalyptic human viruses (Collier et al., 2004). Therefore three areas of concern mark national biosecurity regimes to a greater or lesser extent: (i) the protection of indigenous biota, (ii) agricultural assemblages and (iii) human health. This book encompasses consideration of each of these areas, drawing on a range of different examples and case studies.

In a national context, biosecurity takes on different forms according to its relative importance in regard to national concerns and vulnerabilities. For example, in America, biosecurity is understood primarily in terms of bioterrorism and laboratory biosafety. By contrast, biosecurity in Australia and New Zealand (as well as numerous island states) is driven by concern for native flora and fauna within an environmental conservation ethic, while in Britain and much of Europe the focus is on concern over agricultural assemblages and agricultural pests and diseases. These generalizations, however, belie important details. While New Zealand certainly has stringent environmental biosecurity controls and centres native nature within its national heritage, this stringency is exceeded by measures to control agricultural pests and diseases that threaten the export base on which the country depends. Furthermore, while bioterrorism has been the issue gripping political concern and driving investment in security technologies in the US, this threat has been mobilized in an attempt to make 'naturally occurring' infectious disease relevant to the agenda of policy-makers, resulting in 'dual use' surveillance technologies developed as ways of responding to both these threats (Fearnley, 2007). Finally, in the UK, the control of ruddy ducks (*Oxyura jamaicensis*), hedgehogs and grey squirrels (*Sciurus carolinensis*) was the subject of public political debate well before foot-and-mouth disease spread pyres of burning carcasses across our countryside.

These categories, of course, also overlap as invasive plants spread plant pathogens, as invasive animals introduce disease to agricultural domestic animals, and as a breach of laboratory biosecurity is dealt with as a bioterrorism event. It may in fact be more pertinent to note the ways in which the meanings of biosecurity shift across different sites and in different argumentative contexts,

Table 1.1 Examples of the range of biosecurity-related issues

Problem	Threat	Example	Vector/pathway	Typical response options
Zoonotic disease	Human health; animal welfare; economy	Swine flu; avian flu; SARS	Industrial agriculture; movement of animals and animal products; international travel (humans as carriers); movement of pathogens from wildlife to domestic animals	Destruction of infected animals; containment actions for human and animal populations; vaccination for human and animal populations
Agricultural pests/diseases of animals	Economy; animal welfare; subsistence farming	Foot-and-mouth disease; bluetongue; TB	Usually unintentional, contaminants with animal material	Screening and interception of pests/diseased animals on imported material; destruction of infected animals; containment actions for human and animal populations; vaccination for animal populations
Agricultural/ forestry pests/ diseases of plants	Economy; subsistence farming	Potato blight (*Phytophthora infestans*)	Usually unintentional, contaminants with plant and soil material; horticultural trade	Embargo on movement, sanitation measures of exported goods; destruction of infected plants

Table 1.1 (Continued)

Problem	Threat	Example	Vector/pathway	Typical response options
Marine and aquatic pests and diseases	Environment; economy; recreation	Zebra mussels; eutrophication of waterbodies; didymo	Accidental, human-assisted movement through ballast water and hull fouling; intentional releases of fish species; spreading contamination of inland water bodies through leisure boats and fishing activities	Sanitary measures; public education; mid-journey replacement of ballast water; chemical sterilization of water bodies
Environmental plant pests and diseases	Environment; environmental values; recreational enjoyment of countryside	Sudden oak death (*Phytophthora ramorum*); ash dieback (*Chalara Fraxinea*)	Usually unintentional, contaminants with plant and soil material; horticultural trade	Embargo on movement, sanitation measures for exported goods; destruction of infected plants
Invasive plants and animals	Environment; environmental values; recreational enjoyment of countryside	Possums; rhododendron; Himalayan balsam (*Impatiens glandulifera*)	Accidental as well as intentional human-assisted movement (incl. acclimatization practices, biocontrol; gardening and horticulture)	Intentional: import risk assessments; unintentional: border control/incursion response/pest management
Bioterrorism	Human health; economy; public fear/panic	Anthrax	Intentional release	Preparedness; surveillance; contingency planning; scenario-modelling
Laboratory biosafety	Human health; economy	Foot-and-mouth disease	Accidental release, or bioterrorism act (see above)	Laboratory security; controls on research publication

beyond these broader categories of 'human health', 'agriculture' or 'environment'. In some ways more novel discursively than the practices it denotes (Donaldson et al., 2006), biosecurity can refer simultaneously to the mundane and the extra-ordinary, from precautions such as hand-washing and disinfection, to the spatial management of interactions between people, domestic animals and wildlife, surveillance webs of biocontrol and appropriate paper trails. What draws these different practices and concerns together is a shared construction of threat, posed by the 'dangerous' biological mobility of pests, viruses and other pathogens (Stasiulis, 2004).

In this interdisciplinary edited collection, we approach biosecurity threats as potential biological events precipitated, mediated, made visible, interpreted, politicized and brought into the realm of significance by social, cultural, economic and political factors (Wilkinson et al., 2011). Biosecurity discourses and practices themselves are not simply a response to disease or invasion events, but part of the process through which they are problematized and significance is brought to bear on occurrences. This perspective emphasizes the co-production of disease/invasion and biosecurity policy response, as Stephen Hinchliffe argues in this collection: disease and pest invasions themselves emerge as relational achievements, 'pathogenic entanglements' between environments, human and nonhuman agencies. Importantly, this approach does not belittle the sometimes terrible reality of disease and invasion events – as crops are devastated, livelihoods are ruined and families are bereaved.

Biosecurity has risen to prominence in recent decades due to overlapping security concerns, new global frameworks for managing disease risk, which impact on trade and exports, and the accelerating and intensifying affects of globalization. It is currently high on the political and media agenda, not least following a series of 'focusing events' from SARS to H1N1, West Nile virus to foot-and-mouth, and through the ever present threat of a global outbreak of avian influenza. This concern is materialized in the allocation of greater resources for biosecurity, such as new integrated surveillance systems and networks of laboratories, and global agreements including the International Health Regulations (IHR) in the context of human disease. Are these biosecurity threats growing, alongside the increasingly high-pitched political and media concern? In one domain of biosecurity, Waage and Mumford (2008) analysed trends in trade and agro-ecosystem susceptibility and predicted increasing rates of establishment of new agricultural pests, but pointed to the limited evidence base due to a lack of research. They emphasize, however, that biosecurity burdens are inevitably growing overall, as new introductions accumulate.

The securitization paradigm

Many of the practices that contribute to biosecurity, as well as the threats from pests and diseases themselves, are far from being historically novel. So what marks out contemporary 'biosecurity' as a discursive practice affecting the ways in which the management of unwanted biological mobility – across agriculture, health and environmental management – is imagined, justified and conducted? The securitization of the *bios* has become the dominant response to uncertainty, globalization, rapid mobility and circulatory crises, and terrorism and insecurity. 'Securitization' practices in a host of different contexts entail 'border controls, regimes of surveillance and monitoring, novel forms of individuation and identification … preventative detention or exclusion of those thought to pose significant risks [human and nonhuman], massive investment in the security apparatus and much more' (Lentzos and Rose, 2009, p. 231). Critically, the governance of the future through a regime of uncertainty, urgency and threat is a distinguishing feature of securitization (Caduff, 2008; Anderson, 2010a, 2010b), allowing governance through states of 'insecurity' (Lentzos and Rose, 2009; Lo Yuk-Ping and Thomas, 2010; Brown, 2011).

Biosecurity issues have traditionally been analysed and administered through the lens of risk. Identifying and selecting between risky movements, and determining the level of appropriate intervention and investment in biosecurity measures, involves an industry of risk assessments, risk-profiling, cost-benefit analysis and bio-economic modelling. However, as risk analysis alone is no longer seen as an adequate way of responding to future unknowable events, biosecurity increasingly involves governing through uncertainty and insecurity. Uncertainty, defined by unknowable parameters, is inherent to responses to biosecurity threats and embedded in different biosecurity practices (Fish et al., 2011).

The cloak of uncertainty shrouding biosecurity threats shows itself in the condition of biological emergence, dynamism and indeterminacy (Hinchliffe, 2001; Clark, 2002; Cooper, 2006; Dillon and Lobo-Guerrero, 2009). Rather than a knowable list of historic pests and diseases, unruly biological life has come to be understood as emergent phenomena, presenting the continual possibility of new, unknown and unpredicted infectious diseases and invasive pests. An emergent threat, according to Cooper (2006, p. 124), is a threat whose actual occurrence remains irreducibly speculative, impossible to locate or predict, yet always imminent, the 'not if, but when'. This 'emergency of emergence' (Dillon and Reid, 2009) focuses security attention on biological emergence itself, producing a state of permanent warfare against microbes (Lakoff, 2008a), or the targeting of ever-earlier points of intervention in the production of pests – such as the promotion of sterile garden plant varieties. This anticipation of species that have not yet materialized produces a transformed relationship to an unpredictable future which, as Anderson (2010a) highlights, both exceeds our present knowledge and disallows perfect knowledge – the future will surprise and shock.

Practices of governing in the face of uncertainty have now combined with traditional risk management techniques as part of the biosecurity lexicon. These practices include techniques through which future events are rendered thinkable and constituted as problems (Lentzos, 2006; Collier and Lakoff, 2008), technologies of futurity (Fisher and Monahan, 2011) and future-orientated institutional architectures of contingency planning, precaution, preparedness and pre-emption (Lentzos and Rose, 2009). This produces a realignment of the temporal and spatial scales of governance (Collier and Lakoff, 2008), a hastening of the adoption of new forms of surveillance and information management (Fearnley, 2008a; French, 2009; Parry, 2012), and the justification of new modes and technologies of intervention and containment.

'Bio-securitization' is also intimately linked to problems of circulation (Barker, in prep.). While trade and travel and the pathways they offer to pests, diseases and unwanted species have long existed, it is the speed and volume of this traffic that enhances the possibility for new mutations and emergent threats to occur and for regionalized threats to be globally distributed. These human-induced circulations are matched by biological life's own endless capacity for mobility and mutation (Clark, 2002). It is the potential within these circulations for rapid acceleration or amplification through the increasingly complex globalized circulations of people and things, for a crisis of circulation, for perpetual escalation, which distinguishes circulations requiring a security response from other circulatory threats, for example from road traffic accidents. A threatening circulatory crisis undermines normal risk management, avoids containment within acceptable limits and is marked by its widespread disruptive influence (Beck, 1992; Elbe, 2009; Dillon and Lobo-Guerro, 2008). Significantly, this circulatory threat is not simply produced through the biological enhancing capacities of globalization, but itself produces and requires corresponding security-related circulations. These include those of bio-information within surveillance networks, capital through the expanding biosecurity industry and knowledge and expertise through the international capacity-building of biosecurity regimes (Barker, in prep.).

Description of 'sites' of biosecurity practice

Biosecurity practice can be categorized in a number of different ways. One way to understand the breadth of practice that makes up biosecurity regimes is to view practice and policies that act in and on different sites spatially arranged in relative proximity/distance from national territorial borders (see Table 1.2). In this way biosecurity practice can be mapped from international policy-setting to sophisticated border control; from incursion investigations to routine pest management; and from expert interventions to the activities of individuals in the domestic sphere and in their local environment. At each of these sites the differing focus and the interaction of policies and governing institutions produce very specific cultures of practice. Visualizing biosecurity practices in this way also allows us to distinguish between biosecurity 'acts' and biosecurity 'events' (Donaldson, 2008). While the latter receives social, political, media and academic attention, it is the everyday, mundane and routine biosecurity acts that comprise much of the ongoing processes of biosecuring.

Table 1.2 Typical activities at different 'sites' of biosecurity practice

	Biosecurity site	Typical practices	Focus
International	Pre-border or pre-entry	Inspection regimes at departure ports/airports; import bans or pre-shipment import health standards; international surveillance and report systems; education/communication programmes.	Acknowledging potential threat and predicting or preparing for outbreaks. Preventing occurrence of pest or disease through pre-emptive or precautionary action.Capacity-building, contingency planning, scenario-modelling.
	Border or point of entry	Surveillance and inspection of people and goods: aircraft, cargo, mail, passengers and crew, and sea vessels for unintentional and intentional infringements.	Prevention and enforcement.
	Post-border or post-entry quarantine and surveillance	Active, passive and pathway surveillance.	Prevention and rapid response.
	Incursion response	Immediate deployment of biosecurity service provider to undertake eradication or vaccination programme, or the gathering of further information on the species and its distribution. Containment measures in the context of human disease.	Rapid response for eradication or mitigation.
	Pest and disease management	Site- or pest-specific management programmes (removal, control). Public education. Legislative provisions for managing endemic disease. Veterinary checks, laboratory testing, uptake of farm-level biosecurity.	Population suppression to minimize impact. Managing long-term wider impacts of disease burden.

Source: Fish et al. (2011); Barker (2008b)

The pre-border domain

The pre-border policy and management arena is heavily involved in the formation of and adherence to international legislation. The increasing internationalization of biosecurity governance has raised questions for some commentators about a new 'conformist paradigm'. The World Trade Organization's (WTO) 'Sanitary and Phyto-Sanitary (SPS) Agreement' allows for quarantine as a justifiable non-tariff trade barrier. To prevent countries utilizing biosecurity as a disguised restriction on international trade, measures applied have to be based on scientific principles within risk analysis methodologies. The WTO regularly rules on biosecurity-related trade disagreements between countries. Participating countries are also required to notify changes in the occurrence or distribution of pests and diseases in their national environment. In the case of major disease in wildlife, foot-and-mouth disease for example, this will usually result in trade suspension and other management changes, as countries adjust their pre- and post-border controls. Other multilateral agreements with biosecurity implications include the Convention on Biological Diversity, which states that 'Each contracting party shall, as far as possible and as appropriate: … Prevent the introduction of, control or eradicate those alien species which threaten ecosystems, habitats or species' (IUCN, 2000, article 8(h)); and the International Health Regulations (IHRs), which require states to notify about any event occurring in their territory that may constitute a 'public health emergency of international concern' (World Health Organization, 2008, p. 2). A number of intergovernmental biosecurity networks operate to develop and adopt standards that can be applied at the national level and to administer notification requirements, including the Food and Agriculture Organization of the UN (FAO), the International Plant Protection Convention (IPPC) within the FAO and the World Organisation for Animal Health (OIE).

The pre-border domain has become increasingly interventionist as states attempt to shift the risk of biosecurity off-shore. Activities include pre-checking of goods and passengers by qualified inspectors in departure ports and airports, developing import requirements, screening import applications, developing risk methodologies for the testing of risk products or pathways, and pre-departure education and communication programmes. A state may have bilateral agreements with a number of its import countries determining pre-border quarantine measures. These tend to be developed on a case-response basis. In addition to formal agreements, bilateral exchanges of information, practices and advice are also significant. As countries attempt to manage risks, biosecurity approaches have been adopted and adapted from one governing context to the next.

Passenger and goods border control

Border control is usually overseen by national trade, agricultural or environmental agencies that provide inspection and oversight of the five different incoming sources of people and goods: aircraft, cargo, mail, passengers and crew, and sea vessels. This is undertaken to screen for unintentional incursions – hitchhikers in passengers' luggage or imported goods, or diseased or infested biological material – and to attempt to find deliberately smuggled items.

Passing through quarantine and inspection services at ports and airports during an overseas holiday is likely to be the most tangible individual experience of biosecurity regulations. It forms the subject of sensational documentary series including *Border Patrol* (New Zealand) and *Border Security* (Australia, USA, Canada), which bring biosecurity awareness, and prejudices, to a wider popular audience. Passengers may be required to declare any activities, including visiting a farm or camping, which may lead to an increased risk of introduction of an unwanted organism. Muddy boots, tents and clothing that may harbour seeds, plant fragments or insects may be checked and cleaned. Detector dog teams are frequently used to patrol lines of passengers. In New Zealand, beagles, seen as the 'friendly face' of the Ministry for Primary Industries, are stationed at international airports and mail centres, and are trained to sniff out biological material. In terms of imported goods, these may be inspected on arrival by biosecurity personnel, or the paperwork accompanying them checked for compliance. Mail may also be X-rayed or inspected. Despite these measures, smuggling does occur due to the inevitable permeability of the border, with outbreaks of foot-and-mouth in the UK in 2001, and bird flu in Egypt in 2004, attributed to illegal imports.

Post-border: surveillance

Biosurveillance, defined as the production, analysis and circulation of information on potential invasive events or epidemics, is a crucial ongoing aspect of biosecurity practice. There are a plethora of different forms of biosecurity surveillance at work in and across different sites, that detect new incursions or oncoming epidemics and monitor the health and 'pest status' of plants, animals and ecosystems or progression of a virus during an epidemic. Surveillance networks are not automated information exchanges, but mixtures of humans, nonhumans and technologies performing different practices, including visual inspections, counting, photographing, reporting, sniffing, X-raying, measuring, swabbing, weighing, scanning, recording, collecting and sampling – practices through which biological markers are transformed and circulated as information (Barker, in prep.). These systems are an attempt at calculability, at prediction, at anticipating epidemics or invasive events, allowing rapid response and reconstruction. In the domain of health security, a host of different practices and technologies are being drawn into the health surveillance net, with data on pharmaceutical sales and electronic information generated through the internet (Google searches, social media status updates or automatic searches of news stories) being used to provide early warnings for oncoming epidemics.

Biosecurity surveillance activities can be categorized according to different types of surveillance practice. Systematic and routine 'active surveillance' is made up of targeted surveillance programmes that involve looking for specific organisms in specified 'places' (geographic locations or host species), such as surveys of national agricultural systems for new pests and diseases. This includes 'pathway surveillance', which targets high-risk sites attached to specific risk pathways to look for unspecific risk organisms that may be gaining entry to a country, and feeding back to tighten those pathways. For the second key mode of surveillance, 'passive surveillance', neither the population or territory at risk, nor the unwanted entity, is defined in advance. Passive surveillance can be undertaken by both biosecurity or industry experts, or the general public, through a heightened general 'watchfulness', and involves investigating possible sightings of unwanted organisms.

Post-border: incursion response and pest management

After the detection of an unwanted organism, pest or disease through surveillance mechanisms, rapid response to the incursion should be mobilized, as lower costs and greater capacity for eradication are evident at earlier stages of pest and disease establishment. Attempts are made to identify the organism and its current distribution, before management options are assessed. Depending on the assessed level of risk, this could lead to the immediate deployment of an incursion investigator, diagnostician or quarantine inspector into the field, the deployment of biosecurity service providers to undertake the eradication programme, or the gathering of further information on the species and its distribution.

Pest management is the stage of biosecurity activities that occurs after it is determined that full-scale immediate eradication is not possible due to the extent of a pest's incursion. While activities mentioned above focus on the prevention of new pests and diseases entering and achieving widespread distribution in a country, pest management focuses on recent as well as endemic pests and diseases. The aim of pest management may be to contain the species to prevent its spread to other unaffected parts of the country, or reducing the negative impact of the pest. Eradication may still be an aim; however the difference is the envisaged timeframe. Pest management can make up over half of the total expenditure on biosecurity activities. A further significant realm of internal pest management is public education. This varies from talks to interest groups, posters and leaflets, and stalls at relevant events, to television series and commercials. Pest management may be organized, funded and undertaken by official biosecurity personnel, by contractors, by industry groups or by private landowners, depending on different countries' cost recovery models and understandings of the public good.

Introduction to this book

This edited collection draws together contributions from leading scholars across the rich and varied field of biosecurity studies, combining commentary on actual practices and policies with critical analysis of a range of theoretical issues. It is divided into four interdisciplinary parts. Part I, 'Framing biosecurity', sets the context of the book and investigates why we might need biosecurity, as well as its meanings and some of its potential implications. Part II, 'Implementing biosecurity', investigates the various frameworks that underpin biosecurity practices. Part III, 'Biosecurity and geopolitics', deals with the international dimensions of biosecurity, while Part IV on 'Transgressing biosecurity' looks at biosecurity from the point of view of the human/nonhuman relationship and considers the implications of climate change for the way we view biosecurity.

In Chapter 2 Daniel Simberloff makes the case for containment. He begins by offering an extensive list of examples of damage that invasive species can do, ranging from effects on one type or class of native species to the disturbance of entire ecosystems. This can happen through introduced parasites, diseases and invasive predators, and Simberloff stresses the unpredictable nature of the effects of invasion. These uncertainties can be exacerbated by the time lag issue (a problem can remain dormant and then 'explode' when the appropriate catalyst is present) and the way in which combinations of two of more introduced species can create disproportionate damage. Simberloff points out that sometimes introduced species can benefit conservation, but the results of these interventions are themselves unpredictable and the consequences are not always as intended. Uncontained breaches of biosecurity can have severe economic impacts, especially in the agricultural sector (though Simberloff points out that eight out of the nine major food crops in the USA are introduced). Costs and benefits in this area are notoriously hard to quantify as there is no market in species so they have no price; as Simberloff asks, rhetorically, 'What is the economic cost of a conifer invasion to a native southern beech forest?' A further problem is that values may clash: an introduced species may be of great value to hunters and/or to the tourist industry, but also likely to cause significant ecological damage. How are these to be weighed against one another?

In addition, when considering the risks associated with biosecurity and therefore the case for containment, we need to bear in mind the history of human and animal health impacts caused by bio-insecurity. Simberloff offers, among others, the examples of smallpox, syphilis, yellow fever, and the disastrous pandemic of the fourteenth century for humans, and monkey pox and the rinderpest virus for nonhumans. Given the long list of the costs of bio-insecurity discussed in this chapter, the case for containment looks strong, but can it be achieved? Simberloff argues that containment efforts take place in two kinds of context: planned and unplanned breaches of biosecurity. Evidently containment is more likely to succeed in the former case, although even then the problems that dog both categories are present: the unpredictable ways in which species interact, disperse and evolve. Simberloff is critical of the protocols that presently govern the assessment of the likely impacts of breaches of biosecurity around the world, but argues that containment can work, and would work much better if the regulations in containment regimes were considerably tightened up.

Bruce Braun points out in his chapter on the biopolitics of biosecurity (Chapter 3) that we should be talking about biosecurities (plural) rather than biosecurity (singular), since the managing of biological risk takes many different forms in many different contexts. His intention is not to explicate or discuss this range of meanings, but to show how biosecurity brings life into the realm of politics both through regimes of governmentality and – more provocatively – by determining which forms of life will be protected and which will not. This, says Braun, is biopolitics as 'thanatology': an unavoidable consequence of the project of biose-curity is to 'cut' life up into desirable and undesirable forms. This, as Braun puts it, is to insist that, 'biosecurity be read as a political and ethical issue, and not merely a technical or logistical one'. He suggests that this is fundamentally an anthropo-centric politics, in which nonhuman animal life is sacrificed so that human life might survive. Biosecurity is a quotidian as well as an exotic practice (e.g. hand sanitizers in the workplace on the one hand, and Highly Pathogenic Avian Influenza on the other), yet what these practices all have in common is a set of 'ontological presuppositions' about the nature of networked biological life, a series of knowledges and means for assessing and managing biological risks and decisions about what is to live and what is to die. Together, they present the world as 'seething' with biological threats that can never be fully contained, but only vigilantly managed.

Human and nonhuman animal life is interwoven, writes Braun, and processes of globalization have compressed both time and space to the point where the distant other, whose impact on us is located in some indeterminate future, becomes the local present. How are the indeterminacies associated with a world 'overfull' with life to be managed, especially when the past cannot be guaranteed as a reliable guide to the future, asks Braun. He points us towards knowledge practices, such as health surveillance aimed at monitoring viral mutations, and to computer-based scenario-planning designed to map consequences in a large array of conditions and circumstances. Braun concludes by reaffirming the unavoidability of ethical ques-tions at the heart of biosecurity practices: what should live and what should die – and who decides?

Andrew Donaldson opens Part II, on 'Implementing biosecurity', with his chapter on governing biosecurity (Chapter 4). Donaldson uses the UK and New Zealand to illustrate contrasting ways of dealing with biosecurity at the level of national and sub-national institutions. The UK adopts what he refers to as a 'traditional' or 'sectoral' approach in which the agencies and institutions that deal with biosecurity issues often predate the recognition and naming of the issue and are therefore fragmented in relationship to it. UK policy-making in this context tends to be incremental, and the institutional complexities internal to the UK are magnified by the country's relationship with the European Union and other supranational bodies. Donaldson comments that another effect of this incremental approach to biosecurity in the UK is the dominance of animal health over plant health, due to the historical strength of the former in comparison with the latter. This leaves the UK's biosecurity decision-making community less than ideally placed to deal with plant-related challenges.

In contrast, New Zealand practises an 'integrated' approach to biosecurity in which the practice is regarded as a subset of national security. The institutional framework for this approach amounts to a 'biosecurity system', writes Donaldson, that is characterized by overlapping geographical zones of activity aimed at the prevention of breaches of biosecurity (rather than a post-hoc 'cure') through monitoring, licensing and strict border controls. Donaldson comments that even a system of governance as integrated as New Zealand's is still not as integrated as norms in other jurisdictions (e.g. the European Union) suggest it could be. Effective biosecurity practices depend on regimes of surveillance that also have political implications in regard to otherwise taken-for-granted social norms, such as freedom of action and privacy. Like Simberloff, Donaldson confirms that risk analysis lies at the heart of biosecurity, and he, too, points out how difficult risk assessments are in this context. Not only are knowledge claims the subject of dispute, though – there is an ineluctable ethical dimension to take into account too: risky for whom or for what? This is a question for which science on its own cannot have an answer because of its ethical content.

Donaldson also considers the question of who pays for biosecurity measures, and even here the broader political context must be taken into analytical account for a full picture to emerge. In both the UK and New Zealand there is a commitment to sharing the cost between government and producers, but the legitimating rationale is different in each case. In the UK, the policy is presented as a matter of cutting costs, while in New Zealand it is more focused on sharing resources with a view to achieving a greater, and common, good. Donaldson concludes by arguing for a broader understanding of biosecurity beyond a determination to protect health and trade towards the much bigger question of 'how we live in the world and manage our relationships with other species'.

Opi Outhwaite in Chapter 5 deals with the legal frameworks for biosecurity, and many of the problems that Donaldson discusses in his chapter on governing biosecurity recur here. Outhwaite points out that legislation in this context has a history, even if 'biosecurity' was not the formal subject of these regulatory measures (as it had not been 'named'). In the fields of public health and agriculture it is possible to identify legislation from the seventeenth and eighteenth centuries that was about biosecurity in all but name. Outhwaite points to the sectoral and piecemeal nature of this legislation, which has had an effect on the nature of regulation to this day – a point echoed by Donaldson in the previous chapter. She refers to measures that can be taken at three stages of the invasion process, pre-entry, point-of-entry and post-entry, and comments that laws relating to each of these stages are essential for effective biosecurity measures, since without them activities to secure borders would be the subject of legal challenge.

In large measure, biosecurity as a subject and object of legislation has been fitted into previously existing legislation, including in the international context. This has led to some interesting tensions as the demands of biosecurity have clashed with existing international legal regimes. This is particularly the case, argues Outhwaite, with the World Trade Organization. The WTO was founded to regulate international trade in the context of economic liberalization, and biosecurity measures can often be read as restrictions on trade. Such restrictions have to be justified in terms of risk analysis, but we have seen from previous chapters how problematic this can be. Further challenges identified by Outhwaite to the development of effective international law around biosecurity include inconsistencies of language ('biosecurity' itself can be carefully defined yet loosely interpreted, and the same is true of associated terms such as 'the precautionary principle' and 'alien species'), the fact that legislation can quickly become out of date in this fast-moving field and the eternal problem of compliance (policing it and ensuring the conditions in which law can be carried out as well as made). Outhwaite's conclusion is that effective biosecurity legislation is hampered by the tension between domestic and international arrangements, by differences in interpretation of key terms and by the never-ending challenge of continuously updating statutes in the light of developing scientific and cultural biosecurity challenges.

The way we conceive these challenges is determined in part by how we understand biosecurity – but who should we call upon for this understanding, especially when it comes to confronting biosecurity challenges? Gareth Enticott and Katy Wilkinson's chapter (Chapter 6) deals with the vital question of whose knowledge is to count. They describe the historical development of top-down regimes of intervention, driven and legitimated by a mainstream view of science and scientific rationality. In the context of animal diseases – the main focus of their chapter – a key factor is the rise of veterinary science and the consequent dominance of policy and practice by veterinarians. The authority of vets, suggest Enticott and Wilkinson, derives from their scientific expertise. But is this kind of knowledge the best or the only sort when it comes to dealing with animal disease? This question is explored through an examination of the presence and outbreaks of two diseases in the UK in recent years: foot-and-mouth disease and bovine tuberculosis (bTB). A fine-grained analysis of the development of these diseases and the on-the-ground response to them shows two things. First, veterinary knowledge, far from being monolithic and securely predictive, is itself fragmented and disputed. Second, in the field it becomes apparent that simple rule-following according to codified practices will not always produce the desired results. Practitioners find themselves having to deal with uncertainty and unpredictability, and in the field this leads to coalescences of practitioners into 'communities of practice' where embodied, localized and intuitive forms of reasoning run alongside the scientific rationality that drives mainstream approaches to animal disease and other biosecurity challenges. In the bTB case there was evidence of a reluctance to follow standard procedures, as best practice based on experience in the field and on knowledge shared among practitioners comes to stand in for codified approaches to dealing with animal disease.

In the light of this analysis, Enticott and Wilkinson suggest a broadening of the biosecurity evidence base. They refer to knowledge that is 'more than scientific', and make a plea for the social sciences to play more of a role in what they call an interdisciplinary approach to animal disease. They acknowledge, though, that this recommendation is more easily made than observed in the breach and they point to counterveiling issues such as the way monodisciplinary research is favoured in national assessments, how building effective interdisciplinary research can be a slow process and how getting this research into policy is difficult given the rapid turnover of civil servants and the consequent presumption in favour of the known over the new.

John Mumford (Chapter 7) discusses the practices followed when determining and delivering a response to potential or actual breaches of biosecurity. He offers a detailed account of the procedures followed in determining the nature and extent of risk and points out that it is the responsibility of domestic agencies to carry out risk assessments. This leaves open the possibility – even the likelihood – of uneven and inconsistent risk assessments that militate against effective biosecurity practices across political borders and jurisdictions. The assessment of risk can relate either to individual species or to pathways along which they might travel, and while the latter offers the possibility of a more comprehensive assessment, it is also more complex and subject to indeterminacies. Mumford discusses the pre-entry and post-entry moments of biosecurity practice, commenting on quarantine and surveillance, eradication and pest management, and illustrating these practices with examples from across the world. In the pre-entry context Mumford points out that the problem of the consistent application of norms across jurisdictions arises once again, as 'equivalent measures' are hard to quantify and codify.

Post-entry, the biosecurity practice options are either eradication- or pest management-based (if eradication is not possible or practical). As other contributors point out, the extent of the action taken depends on a cost-benefit analysis (CBA). Mumford amplifies the discussion of CBA in other chapters by commenting on the significance of establishing spatial and temporal boundaries in determining costs and benefits. The temporal dimension is especially complex as it involves prediction at levels of increasing uncertainty, as well as consideration of the thorny issue of an appropriate discount rate. Continuing Donaldson's discussion of the 'who pays for biosecurity?' question, Mumford points out that, while the standard model is one in which costs are shared between government and industry, there are examples of private individuals and the third sector being involved. Mumford also discusses the important phenomenon of cost-sharing at the regional level, particularly in the European Union.

Part III, 'Biosecurity and geopolitics', starts with Clive Potter's repositioning of the role of the WTO in biosecurity, particularly in the context of what he calls 'neoliberal biosecurity' (Chapter 8). Focusing especially on plants and plant products (and the growing role of the horticultural industry), Potter points to expanding trade volumes that militate against effective border controls (a point we find made in many of the chapters in this book, and especially those by Braun and by Hinchliffe) and the massive commercial interests that make it unlikely that protectionism will ever gain much political traction. The WTO's response to the problem, argues Potter, is very much of a piece with neoliberal responses to any challenge to free trade: a focus on managerial and technocratic solutions organized around risk assessment. These techniques, he suggests, are only ever refined rather than reassessed, and the political and normative dimensions of the approach itself are not called into question. Risk assessment is itself a political tool, says Potter, which is designed to depoliticize the biosecurity issue and to render disputes amenable to purely technocratic solutions. But as we saw in Outhwaite's chapter, the very terms through which disputes are supposed to be resolved are themselves a matter of dispute, and if we add to this – as Potter does – a reassessment of the very framing of the biosecurity problem (for whom or what, and in regard to what principles, are we seeking to biosecure borders?), we can see that neoliberalism's attempt to take the politics out of biosecurity through the application of technocratic procedures is itself a political act. In this context, solutions to the problem of biosecurity are hard to come by. Given the commercial interests at stake, Potter suggests that a reduction in trade volumes is unlikely. This leads to a focus on improving the practices of biosecurity, and Potter offers the European Union's plant health regime as an example of an attempt to reconcile open economic borders with the demands of biosecurity. He concludes by appealing for an opening-up of the black box of technocratic risk assessment so that its contents – and the possibility of a reassessment of the nature and desirability of the international horticultural trade – become a subject of public debate.

Alan Ingram (Chapter 9) deals with the governance of global health in the context of emerging infectious diseases. He argues that this has entailed a move away from state-centric forms of governance so typical of the 'Westphalian' approach to dealing with international issues, towards a problem-solving procedure involving multiple actors whose place and role in the governance picture is a matter of constant contestation. In this process, the politics of power is never very far away, and any aspiration anyone may have to governing global health on the basis of rational scientific principles alone for the benefit of all is far from being fulfilled. The idea of 'the global' and who is to determine what it means and how it is to be enacted is one particular arena of contestation, with tensions emerging among a variety of actors located across and between the global North and the global South.

Ingram uses four epidemics to illustrate the emergence of global health governance and the geopolitical challenges and tensions it has engendered. These are HIV/AIDS, SARS, H5N1 influenza and H1N1 influenza. The emerging governance regime has a number of features, each of which is either shared right across the experience of dealing with these epidemics, or is a key aspect of one or more of them. These features include upgrading monitoring systems, an expectation that states keep the World Health Organization informed about the status of diseases on a specified list and a move towards the conceptualization and management of global health in terms of security, accompanied by the appropriate language (for example 'intelligence' rather than 'information'). Serious tensions have emerged as global health governance regimes have developed, especially over the sharing of knowledge as government and states harbour suspicions about the commercial or even military use to which their 'competitors' might put that knowledge and the degree to which any benefits of resulting research and development would be equally shared. This has brought different legal regimes into conflict, with global North states invoking global health security while certain global South states referred to the Convention on Biological Diversity, which gives sovereign nations control over materials originating in their territory. Ingram concludes by urging us to note the ways in which the rationalities and technologies of global health are inflected and affected by political and economic power. Only once this nettle is grasped will we be able to chart a course towards health for all.

In Chapter 10, Brian Rappert and Filippa Lentzos look at biosecurity from the point of view of bioterrorism. They offer a brief history of the development of bioterrorism, showing how it moved from being associated with the actions of states and their proxies to part of the potential repertoire of terrorists, in the aftermath of the Cold War. The attack on the Twin Towers in 2001 was something of a watershed for the US government in regard to the anticipation and prevention of terrorism and this, combined with the anthrax attacks in the immediate post-9/11 period, led to a focus on biological weapons as the most likely source of danger from terrorists in regard to the range of possibilities that come under the rubric 'weapons of mass destruction'. Despite the fact that there have been only three confirmed cases of bioterror since the 1980s, it has become a major concern of the United Nations too, with an increasingly large number of states committed to taking measures against the possibility of biological attack.

Rappert and Lentzos report how research has played a key role in developing preparedness for bioterror attacks. There has been a massive increase in funding, and they discuss the effectiveness of this spend and suggest that it might not only be of positive benefit, but is perhaps even counter productive. Much of this spend has been on securing laboratories, and there has been a parallel concern at the possibility of research proposals and publications being mined by terrorists looking to develop and deploy biological weapons. Rappert and Lentzos conclude by looking to the future, and they suggest that most of the bioterrorism focus will be on implementing and institutionalizing modes of prevention and response. This, they suggest, will be accompanied by yet more diffusion of the meaning and therefore the institutional reach of bioterrorism, and by an agenda that will continue to be set by reaction to events, such as those at the beginning of the millennium that have done so much to shape responses in recent years.

Opening Part IV, 'Transgressing biosecurity', Juliet Fall in Chapter 11 takes up the challenge of interrogating the meaning and significance of language used in debates around biosecurity. Words like 'alien', 'native' and 'invasive' are obviously heavily freighted, yet there is a curious disconnect between their use in scientific discourse and the meanings they have in the social and political world. Take the sentence 'We prefer natives to aliens' and consider two contexts in which it might be uttered. In the discourse of biosecurity it is an explicable and understandable sentiment, but in the social and political world it is unlikely to be heard outside of meetings of racists and xenophobes. Is this significant, or is it just a discursive curiosity?, asks Fall. In the first place she points out the importance of distinguishing between 'biodiversity' and 'biosecurity', in that the former can be achieved at the expense of the latter. Biodiversity can be enhanced by adding another species to those already found in a given place – but from a biosecurity point of view this importing of an 'unruly' species could have disastrous consequences, as other chapters in this book amply demonstrate. During the nineteenth century, indeed, moving species around the globe was regarded as a good thing, as imported plants confirmed the cosmopolitanism of newly globalizing elites. Fall points out that the vestiges of this impulse remain in the global horticultural industry, the effects of which Clive Potter discusses in his chapter.

By the turn of the twentieth century, the desirability of this movement of plants was being called into question, and responsible behaviour came to be seen as more a matter of subtracting species, according to a developing notion of nativism, than adding them. Fall discusses the effect that language usage has had on debates within the biosecurity community (especially between natural and social scientists) and between that community and the public at large. In this latter context the effect seems to be two-edged: on the one hand invoking the notions of 'invasion' and 'battles' can mobilize an otherwise quiescent public to biosecurity action, but on the other this language hardly encourages the more pacified relationship between human beings and the nonhuman natural world that some would argue is a precondition for a broader sustainability agenda. Fall also comments on the discomfiting alignment between the borders that determine both political and plant 'nativism'. Citizenship for people is defined through the boundaries of the nation state – and it is also the responsibility of the nation state to draw up lists of permitted and proscribed organisms. In an increasingly cosmopolitan world this jurisdictional division of labour could seem inappropriate, and Fall refers to recent fieldwork in Switzerland that suggests respondents adopt a more transgressive approach to the principles of biosecurity than the language of nativism might suggest they would or should.

If biosecurity is about securing boundaries, then what are we to make of a practice that flagrantly disregards them – and not only in space but in time too? This is effectively what 'rewilding' does, and this is the subject of Chapter 12, by Henry Buller. Buller points out that rewilding – like biosecurity itself – is an 'elusive term'. It can refer to any one of a number of practices, either in isolation or in combination: the creation of self-regulating land communities on a very large scale, restoration ecology, the reintroduction of native species and/or de-domestication. Each of these is discussed in detail, and Buller shows how, despite having apparently opposed intentions, biosecurity and rewilding have three commonalities: they are both informed by norms as well as by science, they both rely on a sophisticated apparatus of biocontrol and they both require active intervention. Despite rewilding's connection with the idea of a return to a state of affairs before human impact, this last feature – the need for active intervention – highlights the essential artifice at work in the idea and the practice. The practices of rewilding cover a wide range of possibilities, from the introduction of elephants to Australia, through attempts at populating spaces with Pleistocene megafauna, to repopulating the English countryside with the humble otter. Critically, the normative foundations of rewilding vary by cultural context, and Buller comments on the contrast between the vision of a rural arcadia in the UK and the idea of a presettlement wilderness in the US. This is further evidence of the 'enculturing' of nature in rewilding, driven as it is by notions of nature that – in their cultural construction – can only ever be loosely foundational. Buller concludes by suggesting that the impact of rewilding on biosecurity needs to be judged on a case-by-case basis. There are examples of rewilding that have caused significant damage to existing flora and fauna – the cane toad in Australia and the wolf in France, for instance – but there are also examples indicating biosecurity benefits. Ultimately, what animates rewilding is excitement at what Buller calls a 'revitalized naturalism'. And it is the unpredictability, dynamism and potential boundlessness of this naturalism that makes its relationship with biosecurity so fascinating and potentially troubling.

In Chapter 13, Steve Hinchliffe examines biosecurity form the point of view of human/nonhuman relationships, focusing especially on disease and drawing on recent fieldwork in the UK's poultry industry. He points out that 75 per cent of the 200 or so emerging infectious diseases that have been identified recently involve the potential for transmission from nonhuman animals to humans. In this sense the world of biosecurity is a 'more than human' world in which potential crossings between species are more the norm than the exception. This increase in the number of diseases that can cross from nonhumans to humans is due in part, argues Hinchliffe, to the circulations associated with globalization and to the modernization of livestock farming practices. This shows that biosecurity is about more than biology and that a fuller account of its principles and practices has to take into account animal–human social life – 'political virology' – as well. Hinchliffe examines the importance of this insight in the context of preventative practices of biosecurity in the poultry industry in the UK. These practices take the form of a focus on the isolation and separation of poultry from potential disease threats 'from outside', and Hinchliffe wonders whether biosecurity pursued in this form will not simply create more forms of insecurity. This is because the focus on isolation becomes an assumption of isolation, and a consequent failure to attend (a) to the circulations that still take place and (b) to the potential for the biosecured poultry itself to produce disease. These points are illustrated by material from interviews with vets and other industry experts, and they reveal that the movement of birds in the growing process and minimal staffing in the industry combine to produce the possibility of an increase in the incidence of zoonotic disease. Hinchliffe argues that regarding biosecurity as mere 'enclosure' could result in more insecurity. This is exacerbated by the creation of an agricultural infrastructure designed around enclosure and by the simultaneous dismantling of the agencies of effective public response to the outbreak of disease. His worrying conclusion is that, 'The conditions for emergence, infection and transmission may never have been so favourable.'

In Chapter 14, the concluding chapter, the editors of the book, Sarah Taylor, Andrew Dobson and Kezia Barker, look at the future of biosecurity, with a particular focus on the likely impact of climate change, and summarize the themes of the book with reference to specific chapters. Climate change introduces yet another dynamic element into a situation already characterized by fast-moving change and flux, which presents plants and animals with the challenge of adapting to changing environments at a pace potentially beyond their capacity. At the same time, new combinations of species are possible in climate-changed territories. Under these conditions, adaptive and flexible management regimes could be the most appropriate – and this perhaps runs counter to the 'lockdown' approach associated with some biosecurity practices. The tools for the construction of these regimes are discussed, such as developing technologies of surveillance (remote sensing, digital spatial mapping). One of the themes of the book is the unpredictability of the consequences of flows of life, and climate change – unpredictable itself in nature and consequences – adds to the complication. One constant, though, is the need to keep making biosecurity decisions – what is to live and what is to die – and to find ways of putting these decisions into practice. Climate change policy is increasingly characterized by adaptation rather than mitigation (adapting to changed circumstances rather than trying to avoid them), argue Taylor, Dobson and Barker. In the biosecurity context this suggests strains on the policy of containment, as its potential is undermined by the unpredictable and shifting patterns of organism movement as climate change takes increasing hold.

This chapter on the future of biosecurity in the context of climate change completes what we believe to be a comprehensive introduction to the topic of biosecurity. Our contributors follow biosecurity round the disciplines, places and practices where it is found and enacted. They analyse its meanings and implications, and the normative and practical challenges to which it gives rise. Biosecurity emerges from this book as a test case for one of the biggest challenges facing humanity as the twenty-first century unfolds: how to manage our relationship with the nonhuman natural world. There is no guidebook for meeting this challenge, just a series of signs and signals to analyse and negotiate. Those signs and signals, and their analysis and negotiation, are the subject of the chapters that follow.

References

Ali, H. and Keil, R. (eds) (2008) *Networked disease: emerging infections in the global city*, Wiley-Blackwell, West Sussex.

Anderson, B. (2010a) 'Preemption, precaution, preparedness: anticipatory action and future geographies', *Progress in Human Geography*, vol. 34, no. 6, pp. 777–98.

—— (2010b) 'Security and the future: anticipating the event of terror', *Geoforum*, vol. 41, no. 2, pp. 227–35.

Araújo, M. B. and Peterson, A. T. (2012) 'Uses and misuses of bioclimatic envelope modeling', *Ecology*, vol. 93, no. 7, pp. 1527–39.

Barker, K. (in preparation) 'Biosecuring circulations from the microbe to the macrocosm', Following conference presentation in chair's plenary session '(In)Secure Spaces', RGS–IBG 2012 Annual Conference, Edinburgh.

—— (2010) 'Biosecure citizenship: politicising symbiotic associations and the construction of biological threat', *Transactions of the Association of British Geographers*, vol. 35, pp. 350–63.

—— (2008a) 'Flexible boundaries in biosecurity: accommodating gorse in Aotearoa New Zealand', *Environment and Planning A*, vol. 40, no. 7, pp. 1598–1614.

—— (2008b) 'Cultivating biosecurity: governance, citizenship and gardening in Aotearoa-New Zealand', Ph.D. thesis, University of London.

Beck, U. (1992) *Risk society: towards a new modernity*, Sage, New Delhi.

Bingham, N. and Hinchliffe, S. (2008) 'Mapping the multiplicities of biosecurity', in A. Lakoff and S. J. Collier (eds) *Biosecurity interventions: global health and security in question*, Columbia University Press, New York, pp. 173–94.

Braun, B. (2008) 'Thinking the city through SARS: bodies, topologies, politics', in H. Ali and R. Keil (eds) *Networked disease: emerging infections in the global city*, Wiley-Blackwell, West Sussex.

—— (2007) 'Biopolitics and the molecularisation of life', *Cultural Geographies*, vol. 14, pp. 6–28.

Brown, T. (2011) 'Vulnerability is universal: considering the place of security and vulnerability within contemporary global health discourse', *Social Science and Medicine*, vol. 72, pp. 319–26.

Caduff, C. (2008) 'Anticipations of biosecurity', in A. Lakoff and S. J. Collier (eds) *Biosecurity interventions: global health and security in question*, Columbia University Press, New York, pp. 257–77.

Castree, N. (2003) 'Environmental issues: relational ontologies and hybrid politics', *Progress in Human Geography*, vol. 27, pp. 203–11.

Clark, N. (2002) 'The demon-seed: bioinvasion as the unsettling of environmental cosmopolitanism', *Theory, Culture and Society*, vol. 19, pp. 101–25.

Clayton, N. (2003) 'Weeds, people and contested places', *Environment and History*, vol. 9, pp. 301–31.

Collier, S. and Lakoff, A. (2008) 'The problem of securing health', in A. Lakoff and S. J. Collier (eds) *Biosecurity interventions: global health and security in question*, Columbia University Press, New York, pp. 7–32.

Collier, S. J., Lakoff, A. and Rabinow, P. (2004) 'Biosecurity: towards an anthropology of the contemporary', *Anthropology Today*, vol. 20, pp. 3–7.

Cooper, M. (2006) 'Pre-empting emergence: the biological turn in the war on terror', *Theory, Culture and Society*, vol. 23, no. 4, pp. 113–35.

Dillon, M. and Lobo-Guerrero, L. (2009) 'The biopolitical imaginary of species being', *Theory, Culture and Society*, vol. 26, pp. 11–23.

—— (2008) 'Biopolitics of security in the 21st century: an introduction', *Review of International Studies*, vol. 34, pp. 265–92.

Dillon, M. and Reid, J. (2009) *The liberal way of war: killing to make life live*, Routledge, London.

Donaldson, A. (2008) 'Biosecurity after the event: risk politics and animal disease', *Environment and Planning A*, vol. 40, pp. 1552–67.

Donaldson, A., Lee, R., Ward, N. and Wilkinson, K. (2006) 'Foot and mouth – five years on: the legacy of the 2001 foot and mouth disease crisis for farming and the British countryside', Centre for Rural Economy Discussion Paper 6, Newcastle University.

Donaldson, W. and Wood, D. (2004) 'Surveilling strange materialities: categorization in the evolving geographies of FMD biosecurity', *Environment and Planning D*, vol. 22, pp. 373–91.

Dunlap, T. (1999) *Nature and the English diaspora: environment and history in the United States, Canada, Australia and New Zealand*, Cambridge University Press, Cambridge.

Elbe, S. (2009) *Virus alert: security, governmentality, and the AIDS pandemic*, Columbia University Press, New York.

—— (2005) 'AIDS, security, biopolitics', *International Relations*, vol. 19, pp. 403–19.

Farmer, P. (1999) *Infections and inequalities: the modern plagues*, University of California Press, London.

Fearnley, L. (2008a) 'Signals come and go: syndromic surveillance and styles of biosecurity', *Environment and Planning A*, vol. 40, no. 7, pp. 1615–32.

—— (2008b) 'Redesigning syndromic surveillance for biosecurity', in A. Lakoff and S. J. Collier (eds) *Biosecurity interventions: global health and security in question*, Columbia University Press, New York, pp. 61–88.

Fidler, D. (2004) 'Germs, governance and global public health in the wake of SARS', *The Journal of Clinical Investigation*, vol. 113, no. 6, pp. 799–804.

Fish, R., Austin, Z., Christley, R., Haygarth, P., Heathwaite, L., Latham, S., Medd, W., Mort, M., Oliver, D. M., Pickup, R., Wastling, J. M. and Wynne, B. (2011) 'Uncertainties in the governance of animal disease: an interdisciplinary framework for analysis', *Philosophical Transactions of the Royal Society B*, vol. 366, pp. 2023–34.

Fisher, J. and Monahan, T. (2011) 'The 'biosecuritization' of healthcare delivery: examples of post-9/11 technological imperatives', *Social Science and Medicine*, vol. 72, pp. 545–52.

Foucault, M. (2007) *Security, territory, population: lectures at the Collège de France, 1977–78*, Palgrave, New York.

French, M. A. (2009) 'Woven of war-time fabrics: the globalization of public health surveillance', *Surveillance and Society*, vol. 6, no. 2, pp. 101–15.

Hajer, M. A. (1995) *The politics of environmental discourse: ecological modernization and the policy process*, Oxford University Press, Oxford.

Hinchliffe, S. (2001) 'Indeterminacy in-decisions: science, policy and politics in the BSE (bovine spongiform encephalopathy) crisis', *Transactions of the Institute of British Geographers*, vol. 26, pp. 184–204.

Huey, R. B., Kearney, M. R., Krockenberger, A., Holtum, J. A., Jess, M. and Williams, S. E. (2012) 'Predicting organismal vulnerability to climate warming: roles of behaviour, physiology and adaptation,' *Philosophical Transactions of the Royal Society B*, vol. 367, no. 1596, pp. 1665–79.

Ingram, A. (2010) 'Governmentality and security in the US president's Emergency Plan for AIDS Relief (PEPFAR)', *Geoforum* vol. 41, no. 4, pp. 607–16.

—— (2009) 'The geopolitics of disease', *Geography Compass*, vol. 3, pp. 1–14.

—— (2005) 'The new geopolitics of disease: between global health and global security', *Geopolitics*, vol. 10, no. 3, pp. 522–45.

Irvine, W. and Anderson, A. R. (2005) 'The impacts of foot and mouth disease on a peripheral tourism area: the role and effect of crisis management', *Journal of Travel and Tourism Marketing*, vol. 19, no. 2–3, pp. 47–60.

IUCN (International Union for the Conservation of Nature) (2000) 'IUCN guidelines for the prevention of biodiversity loss caused by alien invasive species', prepared by the SSC Invasive Species Specialist Group, http://cmsdata.iucn.org/downloads/2000_feb_prevention_of_biodiv_loss_invasive_species.pdf, accessed 21 September 2012.

King, N. (2003) 'The influence of anxiety: September 11, bioterrorism, and American public health', *Journal of the History of Medicine*, vol. 58, no. 4, pp. 422–41.

—— (2002) 'Security, disease, commerce: ideologies of postcolonial global health', *Social Studies of Science*, vol. 32, no. 5/6, pp. 763–89.

Lakoff, A. (2008) 'From population to vital system: national security and the changing object of public health', in A. Lakoff and S. J. Collier (eds) *Biosecurity interventions: global health and security in question*, Columbia University Press, New York, pp. 33–60.

Lakoff, A. and Collier, S. (2008) *Biosecurity interventions: global health and security in question*, Columbia University Press, New York.

Latour, Bruno (2004) *Politics of nature: how to bring the sciences into democracy*, Harvard University Press, Cambridge, MA.

Law, J. (2006) 'Disaster in agriculture: or foot and mouth mobilities', *Environment and Planning A*, vol. 38, pp. 227–39.

Lentzos, F. (2006) 'Rationality, risk and response: a research agenda for biosecurity', *Biosocieties*, vol. 1, pp. 453–64.

Lentzos, F. and Rose, N. (2009) 'Governing insecurity: contingency planning, protection, resilience', *Economy and Society*, vol. 38, no. 2, pp. 230–54.

Lo Yuk-Ping, C. and Thomas, N. (2010) 'How is health a security issue? Politics, responses and issues', *Health Policy and Planning*, vol. 25, pp. 447–53.

Mather, C. and Marshall, A. (2011) 'Biosecurity's unruly spaces', *The Geographical Journal*, vol. 177, no. 4, pp. 300–10.

Parry, B. (2012) 'Domesticating bio-surveillance: "containment" and the politics of bioinformation', *Health and Place*, vol. 8, no. 4, pp. 718–25.

Samimian-Damash, L. (2009) 'A pre-event configuration for biological threats: preparedness and the constitution of biosecurity events', *American Ethnologist*, vol. 36, no. 3, pp. 478–91.

Sparke, M. (2009) 'On denationalization as neoliberalisation: biopolitics, class interest and the incompleteness of citizenship', *Political Power and Social Theory*, vol. 20, pp. 287–300.

Stasiulis, D. (2004) 'Hybrid citizenship and what's left', *Citizenship Studies*, vol. 8, pp. 295–303.

Szerszynski, B., Heim, W. and Waterton, C. (2003) *Nature performed: environment, culture and performance*, Blackwell, Oxford.

The Royal Society and Wellcome Trust (2004) 'Do no harm: reducing the potential for the misuse of life science research', Report of a Royal Society–Wellcome Trust meeting held at the Royal Society on 7 October 2004, available at www.wellcome.ac.uk/stellent/groups/corporatesite/@policy_communications/documents/web_document/wtx023408.pdf, accessed 27 September 2012.

Tomlinson, I. and Potter, C. (2010) '"Too little, too late"? science, policy and Dutch elm disease in the UK', *Journal of Historical Geography*, vol. 36, no. 2, pp. 121–31.

Waage, J. and Mumford, J. (2008) 'Agricultural biosecurity', *Philosophical Transactions of the Royal Society B*, vol. 363, no. 1492, pp. 863–76.

Wallace, R. (2009) 'Breeding influenza: the political virology of offshore farming', *Antipode*, vol. 41, no. 5, pp. 916–51.

Whatmore, S. (2002) *Hybrid geographies: natures, cultures, spaces*, Sage, London.

Wilkinson, K., Grant, W., Green, L., Hunter, S., Jeger, M., Lowe, P., Medley, G. F., Mills, P., Phillipson, J., Poppy, G. M. and Waage, J. (2011) 'Infectious diseases of animals and plants: an interdisciplinary approach', *Philosophical Transactions of the Royal Society B*, vol. 366, no. 1573, pp. 1933–42.

World Health Organization (2008) 'International health regulations: guidance for national policy-makers and partners', www.who.int/ihr/lyon/WHO_CDS_EPR_IHR_2007_2EN.pdf, accessed 22 November 2012.

2

A WORLD IN PERIL?

The case for containment

Daniel Simberloff

Introduction

Impacts of introduced species are often highly idiosyncratic and affect biodiversity, economic interests, and human health. Although consequences of most introduced species are unknown, many are known to have drastic impacts. Ominously, introduced species sometimes remain scarce and innocuous for decades after establishing populations, then begin to spread and produce major impacts. In addition, consequential impacts of some introduced species are not recognized quickly. Finally, impacts of some introduced species are enhanced after another species is introduced. These facts suggest that even an invasion that is currently apparently harmless may nonetheless warrant concern.

Ecological impacts

Biologists until recently focused mainly on how a particular invasion affects one native species or a class of natives. For instance, many studies have demonstrated devastating impacts of introduced rats on various island seabird populations (Pascal, 2011) and of goats on endemic island vegetation (e.g. Cronk, 1989). The brown tree snake (*Boiga irregularis*), stowing away in cargo shipped from the Admiralty Islands to Guam shortly after World War II, eliminated 15 of 16 native forest bird species (Lockwood et al., 2007), becoming a poster child for the threat of invasive species. More recently, the spread of the introduced Burmese python (*Python molurus bivittatus*) in Florida has generated enormous interest, heightened by recent evidence of a 90+ percent decline in prey species such as raccoons, bobcats, and rabbits (Dorcas et al., 2012). Among notable other destruction caused by introduced predators are the extinction of over 200 endemic fish species by the introduction of the Nile perch (*Lates niloticus*) to Lake Victoria (Pringle, 2011), the extinction of three native fishes and devastation of fisheries after the sea lamprey (*Petromyzon marinus*) reached the North American Great Lakes (Sorensen and Bergstedt, 2011), and the extinction of over 50 endemic snail species on Pacific islands by the rosy wolf snail (*Euglandina rosa*), introduced for biological control of the giant African snail (*Lissachatina fulica*) (Cowie, 2002).

Much recent research in invasion biology has focused on invasions that affect entire ecosystems, 'changing the rules of the game' for many species simultaneously (Simberloff, 2011). Early studies by Peter Vitousek and his associates (Vitousek et al., 1987) on invasion by nitrogen-fixing firebush (*Morella faya*) into nutrient-poor soils of the young volcanic island of Hawaii foreshadowed this trend. The endemic native plants, none of which are nitrogen-fixers, have evolved adaptations to low nitrogen levels that had prevented many ornamental and pasture species planted on the island from spreading. Now, as firebush fertilizes the soil, this constraint is released. Introduction of other nitrogen-fixers into nitrogen-poor habitats has similar effects. Other plants, such as introduced bridal creeper (*Asparagus asparagoides*) in South Australia (Turner et al., 2008), enhance phosphorus soil concentrations and similarly foster invasion by other, previously nutrient-limited plants. The widespread introduction of northern hemisphere conifers into the southern hemisphere for forestry purposes (Simberloff et al., 2010) has far-reaching consequences. These trees produce a more acidic litter with higher concentrations of defensive chemicals than native trees, slowing decomposition rates and affecting many soil species (Dehlin et al., 2008). Ecosystem transformations caused by changes in biogeochemical cycles such as the nitrogen and phosphorus cycles, though far-reaching, are often subtle and not quickly recognized, as replacement of native plants by introduced ones may take a long time.

Several introduced plant species change entire ecosystems by modifying the fire regime. In Florida, Australian paperbark (*Melaleuca quinquenervia*) fosters more frequent and hotter lightning-sparked fires, to which paperbark is adapted and native plants are not. This change in the fire cycle is transforming parts of the 'river of grass', sawgrass (*Cladium jamaicense*) and muhly grass (*Muhlenbergia capillaris*) meadows, into paperbark-dominated forests (Schmitz et al., 1997).

Plants can also modify an entire ecosystem by overgrowth. For instance, on Mediterranean coastal shelves, the Pacific 'killer alga' *Caulerpa taxifolia*, introduced in improperly disposed aquarium contents, and its congener *C. racemosa*, which arrived by unknown means, are overgrowing many seagrass meadows and devastating native animal communities (Klein and Verlaque, 2008).

Animals or diseases that remove the original dominant vegetation can also affect entire ecosystems. Transformation of eastern North American forests over the last 150 years provides several examples. The spread of Asian chestnut blight fungus (*Cryphonectria parasitica*) in eastern North America in the early twentieth century eliminated what had been the dominant tree in many regions, not only extinguishing insects highly adapted to chestnut (Opler, 1978) and modifying nutrient cycling (Ellison et al., 2005), but also bringing many cultural changes to regions closely tied to chestnut products (Freinkel, 2007). The European gypsy moth (*Lymantria dispar*), escaping from an experimental cage in 1868, led to wide-scale replacement of dominant oak species (Liebhold et al., 1995). The Asian balsam woolly adelgid (*Adelges piceae*) has eliminated most Fraser fir (*Abies fraseri*) forests over the last 50 years (Smith and Nicholas, 2000), while the hemlock woolly adelgid (*A. tsugae*) is rapidly destroying eastern hemlock (*Tsuga canadensis*) forests (Ellison et al., 2005), and the Asian emerald ash borer (*Agrilus planipennus*), which arrived in the 1990s, threatens to kill most of the billions of ash (*Fraxinus* spp.) trees in eastern North America (Poland and McCullough, 2006). In each instance, changes in structure, chemistry, and dynamics of an ecosystem wrought by elimination of its dominant trees has far-reaching impacts both above and below ground.

Even invasive predators can produce ecosystem-wide impacts. Rats on offshore islands of New Zealand devastate populations of seabirds that previously nested in great densities in burrows they constructed. By feeding at sea, these birds transfer nutrients from sea to land in the form of food they bring to nestlings, and guano, bird carcasses, and eggs that they leave. Elimination of this transfer, and also absence of burrowing now that seabirds are vanquished, has cascading impacts on plants, below-ground species, and nutrient cycles (Fukami et al., 2006). Introduced foxes preying on seabirds in the Aleutian Islands similarly have effects that cascade through entire ecosystems (Croll et al., 2005), as do introduced yellow crazy ants preying on the red land crabs of Christmas Island (O'Dowd et al., 2003).

In addition to ecosystem-wide impacts, an increasing number of narrowly focused impacts have been documented. Competition for food sometimes plays a role, as with the replacement of the native red squirrel (*Sciurus vulgaris*) in Great Britain by the North American grey squirrel (*S. carolinensis*). The demise of the red squirrel was hastened by squirrel pox (parapoxvirus), which was introduced with the pox-resistant grey squirrel (Rushton et al., 2006). Even unrelated invaders can outcompete native species, as does the introduced Common wasp (*Vespula vulgaris*) in New Zealand, which outcompetes native birds for 'honeydew' produced by insects (Beggs and Wardle, 2006). Herbivory can bring plants to the brink of extinction, as introduced rabbits have done to the endemic 'cabbage' (*Pringlea antiscorbutica*) of Kerguelen Island (Chapuis et al., 2004). Camels that were introduced to Australia for transport in 1840 and released a century later with the advent of motor vehicles, defoliate preferred plants and locally eliminate some species (Edwards et al., 2010).

Introduced parasites and diseases can devastate particular species. Avian malaria, introduced with caged songbirds and vectored by introduced mosquitoes, has contributed to eliminating or threatening most native Hawaiian land birds (van Riper et al., 1986), and crayfish plague (*Aphanomyces astaci*), introduced to Europe with resistant North American crayfish, has facilitated replacement of susceptible European crayfish (Lodge et al., 2012). Dutch elm disease (*Ophiostoma novo-ulmi*) from Asia ravaged both European and American elms in the early twentieth century (Brasier, 2000). Native rainbow trout (*Oncorhynchus mykiss*) in parts of North America have been devastated by whirling disease, brought from Europe in frozen introduced rainbow trout (Bartholomew and Reno, 2002), while many North American bat species are now threatened by white-nose disease, a fungus brought from Europe (Hallam and McCracken, 2010).

Many native species now face a 'genetic extinction' through hybridization with more common invaders. Hawaiian native ducks (*Anas wyvilliana*) and New Zealand grey ducks (*A. superciliosa superciliosa*) have extensively hybridized with North American mallards (*A. platyrhynchos*) introduced for hunting, creating hybrid swarms (Rhymer and Simberloff, 1996). Hybridization has contributed to 38 percent of all native fish extinctions in North America in the twentieth century (Wilson, 1992). Increasing use of molecular genetic techniques reveals an escalating number of cases of massive hybridization. In Scotland, for instance, in at least one area about 40 percent of what had been considered red deer (*Cervus elaphus*) are actually hybrids with Asian sika deer (*C. nippon*) (Senn and Pemberton, 2009), and about 80 percent of what had been termed wildcats (*Felis silvestris*) in Scotland are hybrids with domestic housecats (Hubbard et al., 1992).

As noted earlier, many idiosyncratic impacts are nonetheless important. For instance, the New World cane toad (*Rhinella marina*), introduced to Australia in a failed attempt to control introduced beetles on sugar cane, has become a popular culture icon of unintended consequences of species introductions (Weber, 2010). Native predators of conservation concern often die when they attack the toxic toad (Smith and Phillips, 2006).

Introduced species can also benefit conservation, for instance by substituting for an extinct species. Introduced honeybees and bumblebees sometimes pollinate plants that had been pollinated by now rare native bees (Goulson, 2003). Even a universally deplored invader, the ship rat, can be locally beneficial – in New Zealand it pollinates some native plants whose pollinators are locally extirpated (Pattemore and Wilcove, 2011). Of course, rats helped extirpate the pollinators, so it would have been far better had they never arrived. Some introduced plants can be used as 'nurse plants' to aid restoration of native species. For example, Caribbean pine (*Pinus caribaea*), invasive elsewhere, aided reestablishment in Sri Lanka of native tree species in a degraded site that the natives could not have colonized without the pines (Ashton et al., 1997). Unfortunately, many more harmful impacts are known than beneficial ones.

Biological control – deliberate introduction of a natural enemy of an invader – was initially used to aid agriculture and forestry, but is increasingly used for conservation purposes. However, some biological control introductions for agriculture have caused great non-target mortality (Simberloff, 2012). The cane toad and rosy wolf snail introductions are examples – neither controlled the target pest and both killed threatened native species. This problem is particularly likely if generalized predators or herbivores are introduced, and these are usually eschewed nowadays by professional practitioners. With species specialized to attack the target pest, however, non-target impacts are much less likely. For instance, in Florida the alligatorweed flea beetle (*Agasicles hygrophila*) successfully controlled aquatic South American alligatorweed (*Alternanthera philoxeroides*) (Center et al., 1997). In certain circumstances, non-target impacts of a generalized feeder may be considered less damaging than the impacts of a targeted invader. Recently, two Asian beetles were released in an effort to control the hemlock woolly adelgid, described above, without substantial host-testing to determine their potential impact on native soft-bodied insects. However, the rapid loss of hemlocks and the unusual habitats they create in eastern forests suggest that, if these beetles stem the invasion (which does not yet seem to be happening), the net ecological effect would be beneficial (Simberloff, 2012) no matter the non-target impacts.

Time lags and invasional meltdown

Often introduced species remain geographically restricted and innocuous for decades or even a century, and then abruptly explode across the landscape (Crooks, 2005). Sometimes this sudden spread follows the arrival of another invader. For example, in Florida the Chinese banyan tree (*Ficus microcarpa*) was a harmless ornamental for decades, unable to reproduce because only the absent fig wasp (*Parapristina verticillata*) can pollinate it (Kauffman et al., 1991). The recent arrival of this wasp has transformed this fig into an invasive threat.

A different mechanism for breaking a time lag occurred in Great Britain for hybrids between native small cordgrass (*Spartina maritima*) and introduced North American smooth cordgrass (*S. alterniflora*). These hybrids were noted throughout the nineteenth century, but could not invade because they were sterile, since chromosomes of the parental species were too dissimilar to allow normal meiosis (i.e. cell division necessary for sexual reproduction). In the late nineteenth century, a mutation occurred in one individual that doubled the chromosome number, creating a new, fertile, and invasive species, common cordgrass (*S. anglica*), because each chromosome could then pair at meiosis with a similar one (Thompson, 1991). Mutations are frequently suggested when a formerly restricted species abruptly spreads, but usually there is no evidence (unlike the cordgrass case). Certainly a mutation can in principle transform a harmless species into a horror. For instance, the killer alga that afflicts Mediterranean coastal shelf areas appears to have undergone a mutation, perhaps in waters near Australia, that made it more cold-tolerant, thus enabling it to withstand the winters of the northern Mediterranean coasts (Famà et al., 2002). The problem is that, for most cases of a terminated lag, no direct evidence exists of a physiological or genetic change that would explain the change.

A more likely explanation is often a subtle environmental change. The complex factors governing population growth of an invader yield a situation in which a minor change in one or more factors can transform an innocuous species into a rapidly spreading one, a phenomenon known as an 'invasion cliff' (Davis, 2009). Thus, the environmental change that breaks a long lag need not be dramatic or even rapid. An example is the spread of Brazilian pepper (*Schinus terebinthifolius*), present in Florida for many decades before spreading to become the most invasive plant in the state. The invasion probably resulted from a lowering of the water table because of agricultural, industrial, and household purposes, as well as soil nutrient increase from fertilizer use and rock-plowing for agriculture (Ewel, 1986). Since many introduced species do not become invasive immediately, this suggests the existence of an 'invasion debt' of impacts that will occur in the future from previous introductions (Essl et al., 2011).

The advent of easily used molecular genetic techniques has shown that some introduced populations have more genetic variation than any native population, because propagules have arrived from several areas (Roman and Darling, 2007). The increased genetic variation can spur invasion by a previously restricted species. New genotypes produced by mixing individuals from formerly disparate populations led to the rapid spread of the long-present brown anole lizard (*Anolis sagrei*) in Florida (Kolbe et al., 2004). The multicolored lady beetle (*Harmonia axyridis*), which had been present in the United States for many years, abruptly became highly invasive in the late 1980s, apparently because of the mixture of different genotypes from eastern and western Asia in the American Midwest. Individuals from this population then spread to Europe, where the species has also become invasive (Lombaert et al., 2010).

The fact that an invasion lag can be broken when a second introduced species arrives, as happened with the fig and its wasp, is a facet of the phenomenon known as 'invasional meltdown', in which two or more introduced species together produce a greater impact than would have been predicted from their individual impacts (Simberloff and Von Holle, 1999). The firebush fertilizing Hawaii by fixing nitrogen is part of a meltdown; it is facilitated by the introduced Japanese white eye (*Zosterops japonicus*), which disperses its seeds (Woodward et al., 1990). In addition, introduced earthworms cluster under firebush and increase the rate of nitrogen addition to the soil (Aplet, 1990), exacerbating the main impact of the plant.

An agricultural invasional meltdown in the United States results from invasion by Asian kudzu (*Pueraria lobata*). Soy, also from Asia and a leading United States crop, is attacked by soybean rust, an Asian fungus that uses kudzu as an alternate host, leading to annual losses of hundreds of millions of dollars (Christiano and Scherm, 2007). Recently the Asian soybean aphid (*Aphis glycines*) has also become a soy pest in the United States, with its impact increased by the presence of common buckthorn (*Rhamnus cathartica*), itself a damaging invasive Old World shrub and an alternate host on which the aphid overwinters – another meltdown (Heimpel et al., 2010).

Economic impacts

The biggest economic impact of introductions is agricultural. Much of this impact is beneficial. For instance, of the nine crops classified as major by the U.S. Department of Agriculture, eight are introduced – all but corn (U.S.D.A., 1997). However, introduced species take a terrible toll on agriculture. In the United States, each year native weeds cause an estimated $28 billion in crop damage, while non-native pathogens cost $23 billion and non-native insects and mites another $16 billion (Pimentel et al., 2000).

Non-native tree species are the basis of many forestry industries, valued at billions of dollars annually – the northern hemisphere conifers introduced to the southern hemisphere are good examples (Simberloff et al., 2010). Against this benefit must be tallied two costs. Ecological costs of invasions by these trees are difficult to tally in economic terms because they usually do not affect markets directly. What is the economic cost of a conifer invasion to a native southern beech forest? The other cost associated with introduced trees and forestry is the often staggering costs of introduced pathogens and insect pests – in the United States alone these are estimated at $4.2 billion annually (Pimentel et al., 2000). Ominously, introduced pathogens and pests are increasingly arriving in areas with forest industries based on non-native species. One example is in Chile, where 2 million hectares of Monterey pine (*Pinus radiata*) plantations are threatened by the recent arrival of a water-mold disease (Durán et al., 2008).

Many introduced fish, mammals, and birds are highly valued by hunters and fishermen, whose expenditures support tourism industries. However, ecological costs of such introductions may be great. For instance, North American rainbow trout and European brown trout (*Salmo trutta*) support a New Zealand sport fishing industry that attracts international tourists. But the trout prey on and compete with native fish, and their predation on native insects has led to nuisance proliferation of aquatic plants (Townsend, 1996). Again, economic costs of these ecological changes are difficult to calculate.

A feature of many cost-benefit analyses of introductions is that the parties reaping benefits differ from those experiencing costs. For example, nurseries profit by selling non-native plants. However, many invasive introduced plants are deliberately introduced for horticulture. All of society bears the costs in terms of lost natural areas, management expenditures, and for some species medical costs from allergic or other reactions. A striking case of benefits and costs accruing to different stakeholders concerns Nile perch introduced to Lake Victoria (Pringle, 2011). Europeans have developed a profitable export industry based on this fish, but local fishermen who previously earned their livelihoods from native fish have been eliminated and regional and local societies have experienced disastrous dislocations (Sauper, 2004).

Human and animal health impacts

Many pandemics originated with introduced pathogens, such as smallpox, mumps, measles, and influenza in the New World and oceanic islands (Van der Weijden et al., 2007) and syphilis in Europe (Quétel, 1986). Often introduced pathogens combine with introduced vectors to wreak havoc on public health, as in the recent arrival of chikungunya in La Réunion and Italy, vectored in both places by the introduced Asian tiger mosquito (*Aedes albopictus*) (Lounibos, 2011). Similarly, yellow fever spread in the New World after the seventeenth-century arrival of slave ships carrying the yellow fever mosquito (*A. aegypti*), while malaria reached the New World with the African malaria mosquito (*Anopheles gambiae*), which appeared in Brazil around 1930 (Lounibos, 2011). The most striking epidemic of all, the fourteenth-century plague pandemic that killed 30 million

Europeans, was triggered by an invasion of infected Asian fleas (Van der Weijden et al., 2007).

Human movement and transport have similarly spread epizootic animal diseases. Perhaps the most devastating case was that of rinderpest virus, brought to Africa in Indian or Arabian cattle in the 1890s. This pathogen ravaged cattle populations but also those of many native animals, such as wildebeest and buffalo, whose populations declined by 95 percent in only 2 years. When their prey populations crashed, carnivore populations also crashed, and humans abandoned large areas (Plowright, 1982). Rinderpest was eradicated in Africa in 2011 after a long campaign (McNeil, 2011). Animal diseases are often introduced by means other than livestock transfer (Hickling, 2011). An example is monkey pox, brought to the United States in Gambian pouched rats (*Cricetomys gambianus*) introduced as pets. Vector-borne diseases may be introduced with vectors, as was West Nile virus (which devastates bird populations in addition to its human impacts), almost certainly brought to the United States in an infected mosquito (Lounibos, 2011).

Predicting introduction impacts

Containment measures must target two different sorts of invasions: those in which an introduction is planned and those resulting from unintended introductions. One might expect fewer planned introductions to result in damaging invasions, because the planning process should have included consideration of possible impacts. However, at least as many planned introductions as unplanned ones lead to invasions (e.g. Gordon and Thomas, 1997).

Part of the reason planned introductions become invasive is that regulations are so lax that invasion possibility need not be seriously considered. A contributing factor is that many invasions have such idiosyncratic indirect impacts that no one would have predicted them. For example, North American red swamp crayfish (*Procambarus clarkia*), brought to Spain as a human food source, were maintained in aquaculture facilities, escaping in 1973. They proliferated, with many harmful population- and ecosystem-level impacts (Rodriguez et al., 2003). However, several predatory wading birds profited from this new prey and increased their population sizes (Tablado et al., 2010). At first this increase was seen as beneficial, but the roosts of these increased wading bird populations produced so much guano, with subsequent impact on the soil, that hundreds of prized ancient cork oaks (*Quercus suber*) died (García et al., 2011).

Although such unpredictable invasion impacts abound, regulation of planned invasions rests largely on risk assessments that try to predict whether a species proposed for import will become invasive. Such assessments are difficult not only because of the contingencies of chains of species interactions, as in the example just given, but because all living organisms have two unpredictable features. They disperse, and they evolve. Nevertheless, a number of risk assessments attempt to estimate probability of invasions.

The Australian Weed Risk Assessment (AWRA) (Pheloung, 2001), which is currently used in Australia and New Zealand and tested elsewhere, uses answers to 49 questions about the history, range, and biology of species proposed for introduction. These are combined to classify proposed introductions as 'likely to invade', 'unlikely to invade', or 'need more information'. Cut-off scores for assigning a species to a category are arbitrarily set, depending on how much risk is acceptable. Post-hoc tests of the AWRA on known invasive species show that most, but not all, invaders would have been classifed as 'likely to invade', the exact fraction determined by cut-off points (e.g. Pheloung, 2001). However, the AWRA would also have misclassified several species that have not become invasive (Smith et al., 1999).

The other main risk assessment approach convenes experts who try to think of all possible risks a planned introduction could carry. By an arbitrary algorithm they classify a proposed introduction as acceptable, unacceptable, or requiring further evidence. The British UK Non-native Organism Risk Assessment, used for any proposed introduction in the United Kingdom as an aid for regulators, is such a system, with each expert estimating the likelihood of an identified risk and the uncertainty he or she feels in estimating that likelihood. Scores by different experts are combined by one algorithm, and the combinations of likelihood and uncertainty are combined by another arbitrary algorithm to determine whether the risk a species poses is negligible, justifiable, or unacceptable (Baker et al., 2008).

Neither the AWRA nor the British system formally accounts for the cost or magnitude of the impact that confers a risk. That is, a 1 percent probability of risk might be acceptable if the damage, should the impact occur, is slight, whereas a 1 percent probability of a catastrophic impact would, one hopes, be unacceptable. To some extent, thresholds established in the AWRA and British procedure account for the magnitude of an impact, but only indirectly. New Zealand's Environmental Risk Management Authority (superseded by the Environmental Protection Authority) arbitrarily combined magnitude and probability to produce four risk categories: insignificant, low, medium, and high (ERMA, 2007).

Aside from the arbitrary combination algorithms and thresholds, a shortcoming of these systems is that different experts assess risk and magnitude differently. For instance, an agricultural agent may see release of a biological control agent as of little risk and potentially high benefit (saving a crop from a pest or weed), while a conservationist might see the risk as high (the agent may attack a threatened non-target species) and the benefit as minor. All this is to say that these risk assessment procedures confer an air of quantitative certainty, but they have limited ability to predict risk and their application entails several qualitative value judgments.

Risk assessment for unintended introductions is even less conclusive than that for planned introductions. For an unplanned introduction, we do not know which species will arrive, so we cannot use species traits and past history to assess risk, as with the tools just discussed. The best that might be done is to consider a pathway (such as cut flowers, untreated timber, or ballast water), try to imagine all species this pathway might introduce, estimate for each the probability that it will be introduced, then apply a species-specific risk assessment such as the AWRA. Such efforts have been undertaken – for example, the U.S.D.A. Forest Service used this approach to assess the risk that forest pests or pathogens would hitchhike on untreated Siberian larch (*Larix sibirica*) logs (U.S.D.A., 1991). The team focused on 36 of 175 known pests of larch in its native range, and for each estimated probabilities of (1) infesting larch in the region of origin, (2) being carried on a transported log, (3) surviving transport, (4) establishing a population after arrival, and (5) spreading. For each probability, possible rankings were low, low-medium, medium, medium-high, and high. Individual rankings were averaged by an arbitrary algorithm, yielding a team ranking. For six species, the team estimated ecological and economic consequences should the species invade. For ecological impacts, the team simply listed potential impacts of each species, but did not attempt to quantify or state how likely each impact was.

There is again an illusion of quantification here, but this arises simply from arbitrarily assigning scores to a series of guesses, then arbitrarily combining the scores. The process forces detailed consideration of many risks, but it can hardly serve as a useful estimate of invasion probability.

Can containment be effective?

One may reasonably ask on grounds of feasibility whether a containment policy is appropriate even if the likely invasion damage far outweighs benefits. Growing international trade and travel will continue to cause invasions no matter how stringent containment policies are, but it is far from clear that containment is futile. For instance, introduced mammals established populations in Europe and New Zealand at similar rates through the nineteenth century, so each region had 35 introduced mammal species in 1900. At this time, public awareness greatly increased in New Zealand and a series of biosecurity measures were enacted. The number of established introduced mammal species has fallen to 31, while over the same period the number has grown to 85 in Europe (Simberloff et al., 2012 submitted).

Because many introductions are planned, major benefit would derive simply from tightening permission regulations and subjecting every proposed introduction to expert review, rather than relying simply on a black list with the default being permission to import, as in the United States (see e.g. Fowler et al., 2007 and also Fall, Chapter 11, this volume). New Zealand subjects every planned introduction to review, after which a species may go on a white list; the default position is not to allow entry. A retrospective examination of the New Zealand Biosecurity Act of 1993, the foundation for that nation's biosecurity policies, suggested it has been quite effective in preventing invasions, while pointing to areas for improvement (Parliamentary Commissioner for the Environment, 2000).

Containing unplanned invasions is technically more difficult, but it is possible if the will is present. In the United States, a nation with great deficiencies in interdiction procedures (General Accounting Office, 2001), between 1984 and

2000, the Department of Agriculture intercepted over 725,000 non-native organisms (McCullough et al., 2006). We do not know how many would have established populations and become invasive, but the interceptions include many species that are substantially damaging when introduced elsewhere and are not in the United States because of these interceptions. One could always argue, correctly, that this effort is inadequate and a number of species slipped by that have become invasive. However, surely the take-home message from this fact, plus the number of interceptions, is that the system should be tightened, not abandoned (General Accounting Office, 2001).

A new containment imperative derives from increasing evidence that new genotypes of established non-native species can foster a new invasion or accelerate one already underway – for instance, the multicolored lady beetle invasion discussed above, and invasions by reed canarygrass (*Phalaris arundinacea*). This implies that benefit can derive from excluding propagules of non-native species even if these are already established. For planned introductions, this imperative suggests vigilance and a white list procedure for new varieties of ornamental plants or crops, as well as new proposed sites of origin for established non-native animals and plants.

References

Aplet, G. H. (1990) 'Alteration of earthworm community biomass by the alien *Myrica faya* in Hawaii', *Oecologia*, vol. 82, pp. 411–416.

Ashton, P. M. S., Gamage, S., Gunatilleke, I. A. U. N. and Gunatilleke, C. V. S. (1997) 'Restoration of a Sri Lankan rainforest: using Caribbean pine *Pinus caribaea* as a nurse for establishing late-successional tree species', *Journal of Applied Ecology*, vol. 34, pp. 915–925.

Baker, R. H. A., Black, R., Copp, G. H., Haysom, K. A., Hulme, P. E. et al. (2008) 'The UK risk assessment scheme for all nonnative species', *Neobiota*, vol. 7, pp. 46–57.

Bartholomew, J. L. and Reno, P. W. (2002) 'The history and dissemination of whirling disease', *American Fisheries Society Symposium*, vol. 29, pp. 3–24.

Beggs, J. R. and Wardle, D. (2006) 'Keystone species: competition for honeydew among exotic and indigenous species', in R. B. Allen and W. G. Lee (eds) *Biological invasions in New Zealand*, Springer, Berlin, pp. 281–294.

Brasier, C. M. (2000) 'Intercontinental spread and continuing evolution of the Dutch elm disease pathogens', in C. P. Dunn (ed.) *The elms: breeding, conservation, and disease management*, Kluwer, Boston M.A., pp. 61–72.

Center, T. D., Frank, J. H. and Dray, F. A., Jr. (1997) 'Biological control', in D. Simberloff, D. C. Schmitz and T. C. Brown (eds) *Strangers in paradise. Impact and management of nonindigenous species in Florida*, Island Press, Washington D.C., pp. 245–263.

Chapuis, J.-L., Frenot, Y. and Lebouvier, M. (2004) 'Recovery of native plant communities after eradication of rabbits from the subantarctic Kerguelen Islands, and the influence of climate change', *Biological Conservation*, vol. 117, pp. 167–179.

Christiano, R. S. C. and Scherm, H. (2007) 'Quantitative aspects of the spread of Asian soybean rust in the southeastern United States, 2005–2006', *Phytopathology*, vol. 97, pp. 1428–1433.

Cowie, R. H. (2002) 'Invertebrate invasions on Pacific islands and the replacement of unique native faunas: a synthesis of the land and freshwater snails', *Biological Invasions*, vol. 3, pp. 119–136.

Croll, D. A., Maron, J. L., Estes, J. A., Danner, E. M. and Byrd, G. V. (2005) 'Introduced predators transform subarctic islands from grassland to tundra', *Science*, vol. 307, pp. 1959–1961.

Cronk, Q. C. B. (1989) 'The past and present vegetation of St Helena', *Journal of Biogeography*, vol. 16, pp. 47–64.

Crooks, J. A. (2005) 'Lag times and exotic species: the ecology and management of biological invasions in slow-motion', *Écoscience*, vol. 12, pp. 316–329.

Davis, M. A. (2009) *Invasion biology*. Oxford University Press, New York.

Dehlin, H., Peltzer, D. A., Allison, V. J., Yeates, G. W., Nilsson, M.-C. and Wardle, D. A. (2008) 'Tree seedling performance and belowground properties in stands of invasive and native tree species', *New Zealand Journal of Ecology*, vol. 32, pp. 67–79.

Dorcas, M. E., Willson, J. D., Reed, R. N., Snow, R. W., Rochford, M. R. et al. (2012) 'Severe mammal declines coincide with proliferation of invasive Burmese pythons in Everglades National Park', *Proceedings of the National Academy of Sciences (USA)*, vol. 109, no. 7, pp. 2418–2422.

Durán, A., Gryzenhout, M., Slippers, B., Ahumada, R., Rotella, A. et al. (2008) '*Phytophthora pinifolia sp. nov.* associated with a serious needle disease of *Pinus radiata* in Chile', *Plant Pathology*, vol. 57, pp. 715–727.

Edwards, G. P., Zeng, B., Saalfeld, W. K. and Vaarzon-Morel, P. (2010) 'Evaluation of the impacts of feral camels', *Rangeland Journal*, vol. 32, pp. 43–54.

Ellison, A. M., Bank, M. S., Clinton, B. D., Colburn, E. A., Elliott, K. et al. (2005) 'Loss of foundation species: consequences for the structure and dynamics of forested ecosystems', *Frontiers in Ecology and the Environment*, vol. 9, pp. 479–486.

ERMA (Environmental Risk Management Authority) New Zealand (2007) 'Environmental Risk Management Authority', available at www.ermanz.govt.nz, accessed 10 February 2009.

Essl, F., Dullinger, S., Rabitsch, W., Hulme, P. E., Hülber, K. et al. (2011) 'Socioeconomic legacy yields an invasion debt', *Proceedings of the National Academy of Sciences (USA)*, vol. 108, pp. 203–207.

Ewel, J. J. (1986) 'Invasibility: lessons from south Florida', in H. A. Mooney and J. A. Drake (eds) *Ecology of Biological Invasions of North America and Hawaii*, Springer, New York, pp. 214–230.

Famà, P., Jousson, O., Zaninetti, L., Meinesz, A., Dini, F. et al. (2002) 'Genetic polymorphism in *Caulerpa taxifolia* (Ulvophyceae) chloroplast DNA revealed by a PCR-based assay of the invasive Mediterranean strain', *Journal of Evolutionary Biology*, vol. 15, pp. 618–624.

Fowler, A. J., Lodge, D. M. and Hsia, J. F. (2007) 'Failure of the Lacey Act to protect US ecosystems against animal invasions', *Frontiers in Ecology and the Environment*, vol. 5, pp. 353–359.

Freinkel, S. (2007) *American chestnut. The life, death, and rebirth of a perfect tree*, University of California Press, Berkeley.

Fukami, T., Wardle, D. A., Bellingham, P. J., Mulder, C. P. H., Towns, D. R. et al. (2006) 'Above- and below-ground impacts of introduced predators in seabird-dominated island ecosystems', *Ecology Letters*, vol. 9, pp. 1299–1307.

García, L. V., Ramo, C., Aponte, C., Moreno, A., Domínguez, M. T. et al. (2011) 'Protected wading bird species threaten relict centenarian cork oaks in a Mediterranean

Biosphere Reserve: a conservation management conflict', *Biological Conservation*, vol. 144, pp. 764–771.

General Accounting Office (2001) *Invasive species: obstacles hinder federal rapid response to growing threat*, United States Government Printing Office, Washington D.C.

Gordon, D. R. and Thomas, K. P. (1997) 'Florida's invasion by non-indigenous plants: history, screening, and regulation', in D. Simberloff, D. C. Schmitz and T. C. Brown (eds) *Strangers in paradise. Impact and management of nonindigenous species in Florida*, Island Press, Washington D.C., pp. 21–38.

Goulson, D. (2003) 'Effects of introduced bees on native ecosystems', *Annual Review of Ecology, Evolution, and Systematics*, vol. 34, pp. 1–26.

Hallam, T. G. and McCracken, G. F. (2010) 'Management of the panzootic white-nose syndrome through culling of bats', *Conservation Biology*, vol. 25, pp. 189–194.

Heimpel, G. E., Frelich, L. E., Landis, D. A., Hopper, K. R., Hoelmer, K. A. et al. (2010) 'European buckthorn and Asian soybean aphid as components of an extensive invasional meltdown in North America', *Biological Invasions*, vol. 12, pp. 2913–2931.

Hickling, G.J. (2011) 'Pathogens, animal', in D. Simberloff and M. Rejmánek (eds) *Encyclopedia of biological invasions*, University of California Press, Berkeley, pp. 510–514.

Hubbard, A. L., McOrist, S., Jones, T. W., Boid, R., Scott, R. and Easterbee, N. (1992) 'Is survival of European wildcats *Felis silvestris* in Britain threatened by interbreeding with domestic cats?', *Biological Conservation*, vol. 61, pp. 203–208.

Kauffman, S., McKey, D. B., Hossaert-McKey, M. and Horvitz, C. C. (1991) 'Adaptations for a two-phase seed dispersal system involving vertebrates and ants in a hemiepiphytic fig (*Ficus microcarpa*: Moraceae)', *American Journal of Botany*, vol. 78, 971–977.

Klein, J. and Verlaque, M. (2008) 'The *Caulerpa racemosa* invasion: a critical review', *Marine Pollution Bulletin*, vol. 56, pp. 205–225.

Kolbe, J. J., Glor, R. E., Schettino, L. R., Lara, A. C., Larson, A. and Losos, J. B. (2004) 'Genetic variation increases during biological invasion by a Cuban lizard', *Nature*, vol. 431, pp. 177–181.

Liebhold, A. M., MacDonald, W. L., Bergdahl, D. and Mastro, V. C. (1995) 'Invasion by exotic forest pests: a threat to forest ecosystems', *Forest Science Monograph* 30, Society of American Foresters, Washington D.C.

Lockwood, J. L., Hoopes, M. F. and Marchetti, M. P. (2007) *Invasion ecology*, Blackwell, Malden M.A.

Lodge, D. M., Deines, A., Gherardi, F., Yeo, D. C. J., Arcella, T. et al. (2012) 'Global introductions of crayfishes: evaluating the impact of species invasions on ecosystem services', *Annual Review of Ecology, Evolution, and Systematics*, vol. 43, pp. 449–472.

Lombaert, E., Guillemaud, T., Cornuet, J.-M., Malausa, T., Facon, B. and Estoup, A. (2010) 'Bridgehead effect in the worldwide invasion of the biocontrol harlequin ladybird', *PLoS ONE* 5(3): e9743. DOI:10.1371/journal.pone.0009743.

Lounibos, L. P. (2011) 'Disease vectors, human', in D. Simberloff and M. Rejmánek (eds) *Encyclopedia of biological invasions*, University of California Press, Berkeley, pp. 150–154.

McCullough, D. G., Work, T. T., Cavey, J. F., Liebhold, A. M. and Marshall, D. (2006) 'Interceptions of nonindigenous plant pests at US ports of entry and border crossings over a 17-year period', *Biological Invasions*, vol. 8, pp. 611–630.

McNeil, D. G., Jr. (2011) 'Rinderpest, scourge of cattle, is vanquished', *New York Times*, 28 June, p. D1, available at www.nytimes.com/2011/06/28/health/28rinderpest.html?page wanted=all, accessed 7 September 2012.

O'Dowd, D. J., Green, P. T. and Lake, P. S. (2003) 'Invasional "meltdown" on an oceanic island', *Ecology Letters*, vol. 6, pp. 812–817.

Opler, P. A. (1978) 'Insects of American chestnut: possible importance and conservation concern', in J. McDonald (ed.) *The American chestnut symposium*, West Virginia University Press, Morgantown, pp. 83–85.

Parliamentary Commissioner for the Environment (New Zealand) (2000) *New Zealand under siege: a review of the management of biosecurity risks to the environment*, Parliamentary Commissioner for the Environment, Wellington.

Pascal, M. (2011) 'Rats', in D. Simberloff and M. Rejmánek (eds) *Encyclopedia of biological invasions*, University of California Press, Berkeley, pp. 571–575.

Pattemore, D. E. and Wilcove, D. S. (2011) 'Invasive rats and California recent colonist birds partially compensate for the loss of endemic New Zealand pollinators', *Proceedings of the Royal Society B: Biological Sciences*, vol. 279, no. 1733, pp. 1597–1605.

Pheloung, P. C. (2001) 'Weed risk assessment for plant introductions to Australia', in R. H. Groves, F. D. Panetta and J. G. Virtue (eds) *Weed risk assessment*, CSIRO Publishing, Collingwood, Australia, pp. 83–92.

Pimentel, D., Lach, L., Zuniga, R. and Morrison, D. (2000) 'Environmental and economic costs of nonindigenous species in the United States', *BioScience*, vol. 50, pp. 53–65.

Plowright, W. (1982) 'The effects of rinderpest and rinderpest control on wildlife in Africa', *Symposia of the Zoological Society of London*, vol. 50, pp. 1–28.

Poland, T. M. and McCullough, D. G. (2006) 'Emerald ash borer: invasion of the urban forest and the threat to North America's ash resource', *Journal of Forestry*, vol. 104, pp. 118–124.

Pringle, R. M. (2011) 'Nile perch', in D. Simberloff and M. Rejmánek (eds) *Encyclopedia of biological invasions*, University of California Press, Berkeley, pp. 484–488.

Quétel, C. (1986) *Le mal de Naples*, Seghers, Paris.

Rhymer, J. and Simberloff, D. (1996) 'Extinction by hybridization and introgression', *Annual Review of Ecology and Systematics*, vol. 27, pp. 83–109.

Rodriguez, C. F., Becares, E. and Fernandez-Alaez, M. (2003) 'Shift from clear to turbid phase in Lake Chozas (NW Spain) due to the introduction of American red swamp crayfish (*Procambarus clarkii*)', *Hydrobiologia*, vol. 506, pp. 421–426.

Roman, J. and Darling, J. A. (2007) 'Paradox lost: genetic diversity and the success of aquatic invasions', *Trends in Ecology and Evolution*, vol. 22, pp. 454–464.

Rushton, S. P., Lurz, P. W. W., Gurnell, J., Nettleton, P., Bruemmer, C. et al. (2006) 'Disease threats posed by alien species: the role of a poxvirus in the decline of the native red squirrel in Britain', *Epidemiology and Infection*, vol. 134, pp. 521–533.

Sauper, H. (2004) *Darwin's nightmare* (film), Mille et Une Productions, Paris.

Schmitz, D. C., Simberloff, D., Hofstetter, R. H., Haller, W. and Sutton, D. (1997) 'The ecological impact of nonindigenous plants', in D. Simberloff, D. C. Schmitz and T. C. Brown (eds) *Strangers in paradise. Impact and management of nonindigenous species in Florida*, Island Press, Washington D.C., pp. 39–61.

Senn, H. V. and Pemberton, J. M. (2009) 'Variable extent of hybridization between invasive sika (*Cervus nippon*) and native red deer (*C. elaphus*) in a small geographical area', *Molecular Ecology*, vol. 18, pp. 862–876.

Simberloff, D. (2011) 'How common are invasion-induced ecosystem impacts?', *Biological Invasions*, vol. 13, pp. 1255–1268.

—— (2012) 'Risks of biological control for conservation purposes', *BioControl*, vol. 57, pp. 263–276.

Simberloff, D. and Von Holle, B. (1999) 'Positive interactions of nonindigenous species: invasional meltdown?', *Biological Invasions* vol. 1, pp. 21–32.

Simberloff, D., Nuñez, M., Ledgard, N. J., Pauchard, A., Richardson, D. M. et al. (2010) 'Spread and impact of introduced conifers in South America: lessons from other southern hemisphere regions'. *Austral Ecology*, vol. 35, pp. 489–504.

Simberloff, D., Martin, J.-L., Genovesi, P., Aronson, J., Courchamp, F. et al. (2012) 'Biological invasions: what's what and the way forward'. Manuscript submitted.

Smith, C. S., Lonsdale, W. M. and Fortune, J. (1999) 'When to ignore advice: invasion predictions and decision theory', *Biological Invasions*, vol. 1, pp. 89–96.

Smith, G. F. and Nicholas, N. S. (2000) 'Size- and age-class distributions of Fraser fir following balsam woolly adelgid infestation', *Canadian Journal of Forest Research*, vol. 30, pp. 948–957.

Smith, J. G. and Phillips, B. L. (2006) 'Toxic tucker: the potential impact of cane toads on Australian reptiles', *Pacific Conservation Biology*, vol. 12, pp. 40–49.

Sorensen, P. W. and Bergstedt, R. A. (2011) 'Sea lamprey', in D. Simberloff and M. Rejmánek (eds) *Encyclopedia of biological invasions*, University of California Press, Berkeley, pp. 619–623.

Tablado, Z., Tella, J. L., Sànchez-Zapata, J. A. and Hiraldo, F. (2010) 'The paradox of the long-term positive effects of a North American crayfish on a European community of predators', *Conservation Biology*, vol. 24, pp. 1230–1238.

Thompson, J. D. (1991) 'The biology of an invasive plant: what makes *Spartina anglica* so successful?', *BioScience*, vol. 41, pp. 393–401.

Townsend, C. R. (1996) 'Invasion biology and ecological impacts of brown trout *Salmo trutta* in New Zealand', *Biological Conservation*, vol. 78, pp. 13–22.

Turner, P. J., Scott, J. K. and Spafford, H. (2008) 'The ecological barriers to the recovery of bridal creeper (*Asparagus asparagoides* (L.) Druce) infested sites: impacts on vegetation and the potential increase in other exotic species', *Austral Ecology*, vol. 33, pp. 713–722.

U.S.D.A. (United States Department of Agriculture) Forest Service (1991) *Pest risk assessment of the importation of larch from Siberia and the Soviet Far East*. U.S.D.A. Forest Service, Washington D.C., Miscellaneous Publication no. 1495.

—— National Agricultural Statistics Service (1997) *Agricultural statistics 1997*, United States Government Printing Office, Washington D.C.

Van der Weijden, W., Leewis, R. and Bol, P. (2007) *Biological globalisation*, KNNV, Utrecht.

van Riper, C. III, van Riper, S. G., Goff, M. L. and Laird, M. (1986) 'The epizootiology and ecological significance of malaria in Hawaiian land birds', *Ecological Monographs*, vol. 56, pp. 327–344.

Vitousek, P. M., Walker, L. R., Whiteaker, L. D., Mueller-Dombois, D. and Matson, P. A. (1987) 'Biological invasion by *Myrica faya* alters ecosystem development in Hawaii', *Science*, vol. 238, pp. 802–804.

Weber, K. (ed.) (2010) *Cane toads and other rogue species*. PublicAffairs (Perseus), New York.

Wilson, E. O. (1992) *The diversity of life*. Harvard University Press, Cambridge M.A.

Woodward, S. A., Vitousek, P. M., Matson, K., Hughes, F., Benvenuto, K. and Matson, P. (1990) 'Use of the exotic tree *Myrica faya* by native and exotic birds in Hawai'i Volcanoes National Park', *Pacific Science*, vol. 44, pp. 88–93.

3

POWER OVER LIFE

Biosecurity as biopolitics

Bruce Braun

Introduction

Since the late 1990s, and in the aftermath of events such as the 2001 anthrax attacks, the 2003 SARS crisis and the simmering problem of Highly Pathogenic Avian Influenza (HPAI), we have witnessed the global proliferation of 'biosecurities'. By biosecurity, I mean those knowledges, techniques, practices and institutions whose concern is to secure valued forms of life from biological risks. By pluralizing the term, I mean to note the heterogeneity of biosecurity practices: it would be a mistake to imagine that all forms of managing biological risks are the same, either historically or geographically, or even that in any given place biosecurity names a single or fixed set of practices. One need only think of the differences between responses to zoonotic diseases like SARS and HPAI, efforts to deal with foodborne illnesses such as e-coli (*Escherichiacoli*) and salmonella, and apparatuses of security around bioterrorism, whether in terms of restricting access to actual biological agents or simply to the knowledge needed to produce old or new pathogens. In some countries, like New Zealand, biosecurity extends to the problem of invasive species and preserving the integrity and persistence of particular ecologies, often related to definitions of the nation and national identity. In many cases these diverse practices overlap – in farms, labs, airports, industry and food service – such that one frequently moves in and between multiple sets of biosecurity practices without being aware of doing so.

Attending to the diversity of biosecurity practices is beyond the scope of this short chapter. What I seek to develop in these pages is a reading of biosecurity in terms of its *biopolitical* dimensions; that is, the ways in which biosecurity practices bring 'life' into the realm of political calculation. As we will see, biosecurity involves *power over life*, and does so through a variety of knowledge practices and logistics. Some of these practices and logistics have to do with the government of our everyday lives as biological beings in contact with other biological beings, in line with what Michel Foucault referred to as 'governmentality' (Foucault, 1991). In such instances the goal is to produce governable bodies, or, better, individuals who relate to themselves as *biological* selves, and do so in particular ways, so as to sustain the vitality of populations as a whole. But arguably biosecurity is at times biopolitical in another, more explicitly 'thanatological' manner, in the sense that in their various efforts to secure a valued form of life, biosecurities abandon or pre-empt *other* existing or potential forms of life. In this latter sense, biosecurities can be seen to invest in life through making cuts into the fabric of life as a whole, whereby those forms of life deserving of protection are separated from those forms that can be sacrificed. Such cuts have geographical and geopolitical dimensions that must be attended to carefully in order to attend to how populations are differentially situated in relation to such practices (Braun, 2007; Sparke and Anguelov, 2011). But as will see, the divisions that biosecurity enacts are not merely divisions within or between human populations, but also divisions between humans and nonhumans, and within nonhuman life more generally, such that some forms of animal and plant life are accorded protection and others not. As such biosecurity involves bringing power to bear on biological life *in general* in order to secure *particular* forms of life.[1]

Attending to the biopolitical dimensions of biosecurity is not a reason to dismiss efforts to manage biological risks for humans, animals or plants. Rather, as I will emphasize throughout, doing so calls attention to the *forms of life* cultivated within its practices as well as the inevitable calculus of life and death at their heart. In other words, attending to biosecurity *qua* biopolitics asks how our existence as biological beings is subject to administration, opens key questions about what forms of life are accorded protection and which are not, and invites us to imagine and debate how diverse lives might be made or allowed to flourish *without* requiring us to foreclose on other possible lives.[2] In short, it insists that biosecurity be read as a political and ethical issue, and not merely a technical or logistical one. Or, more to the point, it alerts us to the fact that in its technical and logistic practices are found a set of pressing and irreducible political and ethical concerns.

Governing unruly assemblages

We do not have to travel far to witness biosecurity practices. It is often said that the site of greatest biological risk is one's own kitchen. With the rising risk of e-coli and salmonella – often linked to industrial agriculture – the kitchen is increasingly a site of protocols around the handling of food, personal hygiene, and the like, which aspiring foodies are continuously encouraged to follow. 'Cross-contamina-tion' has become a household phrase. Or take another example that is ready to hand. In the building where I am writing, hand sanitizer dispensers are located beside every elevator, and faculty, staff and students are encouraged to regularly use them. Installed during the 'swine flu' scare of 2009, they quietly perform a simple function, reminding building occupants of the biological risks of social life and nudging them to comport themselves accordingly. Members of the university community are likewise routinely exhorted to get annual flu shots, as part of the institutions' efforts to limit disruptions to their schedules. Although it is often at extraordinary moments – the SARS epidemic of 2003, intermittent scares over avian influenza (HPAI), swine flu, or the (feared) release of deadly pathogens in bioterrorist attacks – that the question of biosecurity and the presence of biose-curity practices are most immediately evident, biosecurity practices are just as often out of sight, enacted in the hidden spaces of industrial agriculture, buried in bureaucratic activities such as preparedness planning, or expressed in the spatial reorganization of livestock and agricultural workers in far-flung countries as part of the wedding of health, environment and development by international organizations, aid agencies and public health departments.[3]

In what follows, I draw out key elements of how biosecurity 'takes hold' of life so as to make life live. I begin by identifying a set of ontological presuppositions located at the heart of many contemporary biosecurity practices.[4] It is these presuppositions which provide biosecurity practices with their warrant. I follow this with a discussion of particular knowledges and logistics by which present and future biological risks are made calculable and actionable, before returning at the end of the chapter to questions of the administration of life.

Proliferating life: biosecurity's ontologies

To say that biosecurity entails a series of ontological presuppositions is to say that it operates with a particular conception of what life *is*. Following Jane Bennett (2010), we might call these 'ontostories', accounts which, to borrow from Dillon and Reid (2009, p. 77), tell us 'what life is and how it operates', and, conversely, what interrupts, disrupts and corrupts life. Such conceptions are historical rather than universal – tied to particular material conditions in which life comes to be understood in particular ways.[5] They are also consequential, insofar as how life is framed relates directly to what sort of political technologies are brought to bear upon it. This can be illustrated through the example of emerging infectious diseases, a problem that preoccupied health authorities in the early 2000s, and which functioned as a key moment in the emergence and dissemination of biosecurity practices.[6] What was striking about the concern with emerging infec-tious diseases, and to a lesser extent bioterrorism, was the *image of life* that informed how biological life was translated into a realm of imminent threat. For heuristic purposes, we can break this image of life into the following elements: the *emergent* nature of biological life; the *symbiotic* relation between human and nonhuman life; the *spatio-temporalities* of global assemblages; and the problem that each of these presents to the *administration* of life. Let me take each of these in turn, using the example of infectious diseases to explore their significance.

One of the most remarked upon aspects of HPAI and SARS was that each involved the introduction, or potential introduction, of novel pathogens into human populations, a concern that surfaced again with the swine flu (H1N1) scare of 2009. In the case of SARS, the story is usually told of a virus that jumped from animals to humans in the 'wet' markets of Guangdong, China and from there to Hong Kong and on into global networks of trade and travel. While at first linked to civet cats (*Viverridae*), a species farmed for food in China, its origins were eventually traced to horseshoe bats (*Rhinolophus* spp.) which lived in close proximity to the farms in which civet cats were raised. For my purposes, the accuracy of this explanation is of less interest than what captured the attention of media, government and publics in and beyond affected areas: that a virus endemic among animal populations had apparently 'jumped' to human populations and spread (or threatened to spread) rapidly across the globe. In the case of HPAI, the concern was amplified as the ever-present *possibility* of such an event happening, and in important respects it was with HPAI that this new image of life found its purest (and most terrifying) expression. The problem with viruses like HPAI, as stressed by virologists like Robert Webster, and continuously rehearsed in the media, was that they were continuously taking new form through the reassortment of genetic sequences between viruses. Thus, they held the potential of suddenly transforming into highly pathogenic forms that could at any moment jump species and, as important, jump scales (see Webster, 2003).

We can see in this a paradox that I will return to later: that the problem of biosecurity is not just a problem of securing life (in the sense of protecting and preserving life), but rather a problem of securing life against the *proliferation* of life.[7] The problem to which biosecurity responds is thus that of *too much life* – reflected in representations of the biological world as unruly, prolific, mutable, fluid, and the accompanying fear that continuously incubating within life are threats *to* life. As such, life must be secured *against* life. Later we will see why this means that biosecurity can never finally evade the problem of thanatopolitics, the problem of what life must be killed in order that some lives may live more.

What made this spectre of proliferating life so unsettling was a second ontological dimension implicit in how the SARS and HPAI scares were presented, and today increasingly explicit in biosecurity practices: the inseparability of human life from the larger bio-technical assemblages within which human life is enfolded. This came to be understood in two ways. On the one hand, with increased attention to zoonotic diseases and food-borne illnesses it became increasingly difficult to sustain the impression that public life was a purely *social* or *political* affair. Biological life – the lives of animals, plants, viruses, bacteria – was increasingly seen to course *through* social and political life, even to the point of providing the larger 'web of life' within which life itself was possible at all.[8] In a sense, 'life' came to be seen as an effect of 'more-than-human' networks, which were both the basis *for* life and that which exposed life to threats. The same was true of technological objects and networks – airplanes, factory farms, laboratories – which, within global assemblages, came to be seen as actants in their own right. Indeed, the spread of SARS to cities like Toronto and Singapore in 2003 made explicit something that epidemiology and microbiology had maintained for some time: that far from self-contained and discrete entities, bodies are by their nature continuously affected by, and exchanging properties with, myriad other bodies, whether human, animal, plant or machine. Indeed, in important respects this symbiotic 'living together' (Barker, 2010) reversed a trend in the analysis of social and urban life that had for decades understood cities as primarily 'cultural', 'economic' and 'political' spaces, pushing aside or ignoring altogether the myriad nonhumans that co-existed within and helped constitute urban life.[9] With the case of SARS, cities like Toronto came to be understood as complex and unpredictable *biosocial* spaces, in some senses returning to an earlier 'epidemiological' understanding of urban life that had been prevalent in the nineteenth century (see Craddock, 2004; Gandy, 2006). No longer were bodies seen as discrete and autonomous. Nor were nonhumans seen as 'outside' social life: rather, animals and technologies were increasingly seen to be intimately interwoven in the fabric of what had for many years been merely as 'human' collectives. Accordingly, transformations in *non*human life now came to be understood as immediately transformative of social, economic and political life, and vice versa (see Braun, 2008; Mitchell, 2002).

The interwoven nature of human and animal life was also revealed in a further sense: the bio-technical assemblages that constituted the lives of individuals were *global* in scale. With the spread of industrial agriculture, and with the intensification of global transportation networks, human–animal relations in distant locations were suddenly seen as proximate in ways that they had not been before (Davis, 2005; Wallace and Kock, 2012). It was not only that animal life was now seen to be intimately interwoven with human life in places like Toronto, Hong Kong or Singapore (in the form of pets, urban wildlife or food products), but that, in a networked world, animal life in Guangdong, China was directly and intimately interwoven with human life in Toronto (and vice versa). Transformations within the genetic sequences of a virus in one location could be seen as threats in another; likewise human–animal relations in any one place could be seen as a concern to populations in far-flung locations.

With this understanding of networked life a third ontological dimension comes into view: the complex *temporalities* of biosocial life. This too had multiple dimensions. In the case of SARS, for instance, it had first to do with the challenges posed by transcontinental air travel. The common account of the epidemic placed great emphasis on the ability of the virus to enter the biosocial world of the travelling classes, and thus to be transmitted globally to unsuspecting others in mere hours. It is precisely this 'folding' of time and space that rendered human–animal relations 'over there' an immediate concern 'over here'.[10] But the *speed* at which viruses could now travel across the world was not the only concern: of equal concern was that infected individuals could be infectious yet still pre-symptomatic, such that there were few ways to discern the movement of the virus until it was too late. In other words, the SARS virus had its own temporality that complicated the story of bio-technical assemblages. Indeed, the issue here was one of multiple temporalities or what Smith (2003) has helpfully called 'polyrhythmic' space. Assemblages may be complex knots of social, ecological and technical relations, but they do not move at the same speed. The speed of global travel networks, for instance, was far more rapid than the progress of the infection in any single individual, rendering global travel a threat. On the other hand, the speed of immunology was of a different order, intensifying the temporal gap between infection and the appearance of symptoms with a second temporal gap between the emergence of the disease among humans in November 2002 and the first generation of diagnostic tests, which were not available until April 2003 (see Braun, 2008). In short, the time of infection and transmission, the time of global transportation networks and the time of knowledge production were distinct yet folded together into the same event, with immense consequences for the administration of public health.

Summing up, we can say that amid the 'bio' scares of the late 1900s and early 2000s, health officials and government authorities – especially in the global North – were increasingly confronted with an understanding of life as *global*, *networked* and *emergent*. As Eugene Thacker (2005) neatly put it, in the 'spatio-temporal multiplicities' that characterized the globalized world, things were 'continuously churned up', forming 'unexpected combinations'. This view of proliferating life was perhaps best captured by Brian Massumi (2009). With neo-liberal globalization, Massumi explains, we now inhabit a world in which 'life' has come to be understood in terms of 'incipient events'; incubating in the present is the possible future catastrophe. In some respects this view of life is a return to a much older one. Massumi notes that in several of its aspects it closely mirrors what Spinoza in the sixteenth century referred to as *natura naturans*. For Spinoza (1996 [1677]), *natura naturans* named the continuous coming into existence of things and thus signified a *potentiality* that existed within all *actual* worlds, a potentiality that was *determined to be determined* in one way or another. The world as it is determined in any one instant (what Spinoza referred to as *natura naturata*, consisting of concrete 'things', including such things as 'ideas') is thus merely a way-station to further transformations. By this view potentiality is never exhausted in the actual; rather, it coexists with it.

Massumi's emphasis on potentiality (or the 'virtual') rather than actuality (or, better, the *inseparability* of the actual and virtual, which are always co-present) captures well the ontology that informs contemporary biosecurity practices. It is precisely the *possibility* that human–animal–technology assemblages may produce 'unexpected combinations' to which diverse biosecurities seek to respond. For if the world is understood as overfull with potential, then human life is always faced with the spectre of what Massumi calls the 'singular-generic' or 'indiscriminable' threats which cannot be predicted in advance. Significantly, this view of global assemblages in terms of 'indiscriminable' threats extends to bioterrorism too, for it is not only the case that human–nonhuman assemblages continuously form unexpected combinations, but that existing (and new) biological entities can be mobilized or deployed unexpectedly by new actors within the shifting skein of global networks that today comprise political and economic relations. As such, the management of biological risks intersects with other practices of risk management in a world of 'emergent' risks (see Cooper, 2008).[11]

Fielding the incipient event: making biological risk calculable

The *emergent* quality of biological life presents obvious problems for the protection and security of valued forms of life: If the world is overfull with potential, how does one act so as to bring about *one* biosocial future rather than *another*? How does one make possible futures *actionable* in the present? Attending to these questions will enable us to further refine the ways in which biosecurity can be seen to be biopolitical.

Drawing upon arguments made by Stephen Collier (2008), we might say that the problem with *emergent* life is that it gives us an understanding of the future as *discontinuous* with the past. The future event has no historical analogue, and thus the past cannot be a reliable guide to the future. This means that traditional methods of archival-statistical reasoning, in which future events are predicted through statistical norms derived from past records, no longer hold force.[12] Instead, other methods for making the future event calculable must be developed and deployed.

This point can perhaps be taken too far because many biological risks are well known. For instance, food-borne illnesses such as salmonella and e-coli are well understood, and a set of policies govern agriculture and food production so as to limit the possibility of outbreaks of either. But even with these there remains a degree of indeterminacy: as assemblages of industrial agriculture change, new and unexpected points of transmission and contagion emerge, and new forms of bacterial and viral life continue to mutate (Wallace and Kock, 2012). Many other risks remain unknown, or remain 'potential', with the 'when' and 'where' of any outbreak impossible to predict in advance. Rendering such potential events actionable thus requires different methods. Ben Anderson (2010) has suggested that these methods are performative in character, no longer seeking to predict what *will* happen, but instead imagining what *could* potentially happen. Such methods include as a central component the modelling and testing of *scenarios*, in order to see how different futures might play themselves out, and what might be done in the present to either pre-empt or prepare for them.[13] To be sure, past events still have a place in such knowledge practices. The 'Great Influenza' of 1918 for instance, is often taken as a base from which to develop future scenarios around infectious disease.[14] But extrapolating from such past events is never sufficient in itself. This is true not simply because *emergent* biological risks are by definition new rather than simply a repetition of the past, but also because the global *assemblages* within which 'life' is constituted today, and from which biological risks may potentially emerge, are of an entirely different nature than in the past. We no longer live in an age of telegraphs and steamships, nor do we live in a world without virology and immunology.

Today, knowledge practices designed to make future risks calculable take numerous forms. On the one hand, they involve the expansion of health surveillance, not only across human populations but animal populations too, in order to register and identify new viral mutations as they emerge. Such efforts to develop an unending picture of human and animal health have been further facilitated by the rapid expansion of social media and other forms of information sharing that enable health authorities in one place (e.g. World Health Organisation (WHO), Center for Disease Control and Prevention (CDC)) to gather information separately from national authorities that may or may not be open to sharing data. It also involves mapping the genomes of viruses in order to develop a comprehensive database that can be used in the rapid development of vaccines, or, as is now being done, the development of 'pre-pandemic' vaccines that can slow the spread of a disease until a vaccine that protects against the specific pandemic virus is produced. Vaccine adjuvants are one promising line of inquiry, as they have shown cross-reactive protection, but experiments also continue on whole virus vaccines and live attenuated vaccines, among others (CIDRAP, 2007). In each of these instances, the goal of knowledge production is to reduce the temporal 'gap' between the beginning of a possible pandemic and the response of local and global institutions of public health.

An equally important set of knowledge practices has been the modelling of what *could* happen in a pandemic (scenario modelling) and testing the effects of different response strategies. Such models begin with particular assumptions – for instance, about the source of a new biological threat, its communicability and incidence rate, proximity to global transportation networks, everyday activities of individuals at work, home, school or play, and even the movements of wildlife that might transmit the disease. IBM Research, for instance, has developed a 'spatial-temporal epidemiological modeler' available for policy makers and planners who wish to develop their own pandemic models. Likewise, in 2007 the Models of Infectious Disease Agent Study (MIDAS) Network charged three different research teams to develop computer models of a pandemic in a city of about 8.6 million people (similar in size to Chicago). Funded by the National Institute of General Medical Sciences (NIGMS), its goal was to document the effects of different interventions, such as social-distancing measures (e.g. closing schools) antiviral treatments, or the two in combination (Halloran et al., 2008). The advantage of such computer models is that they can be run continuously with different initial parameters, thus generating a vast database of possible events, and an equally large database documenting the effects of different interventions.

As Ben Anderson (2010) explains, what these forms of knowledge do is render the future not merely calculable but actionable. Policy and planning can occur in such a way as to act on such future events before they happen, either through pre-empting them, intervening *before* the events occur, or through forms of preparedness planning that seek to mitigate their effects *after* the onset of the event. Pre-emption, Anderson explains, often involves intervening in the conditions of emergence. Returning to the language of Massumi, we might say that it seeks to determine that which is *determined to be determined*: it is a form of *ontopower*, insofar as it seeks to produce *one* biosocial future rather than another. Preparedness, on the other hand, is power over life in a somewhat different register: seeking to produce 'resilient' social, economic and political collectives that can absorb unexpected biosocial events while retaining their systemic character. Each involves *logistics*, modes of planning and action that seek to secure valued forms of life from life's exuberance and unpredictability.

Biosecurity as biopolitics

While knowledge practices like those just discussed enable life to be brought within forms of political calculation, it is with logistics that we can begin to draw out the *biopolitical* dimensions of biosecurity practices most fully. Preparedness, for instance, seeks to organize life in advance of any event so as to enable the persistence of a particular form of life *after* the event. It pre-orders what comes after. Thus, it does not seek to pre-empt the event, which it sees as inevitable ('not if, but when'), but instead orders flows of people, goods, and information in particular ways, something Julian Reid (2007) refers to as 'logistical life'. At one level this involves developing protocols and training responders to instigate practices that will mitigate the effects of biosocial events on everything from medical facilities and global supply chains to critical infrastructure and resources. Such logistics take as their goal maintaining the biosocial networks that allow life to flourish in the face of the unexpected threats that the same – or similar – networks call forth. But it also involves altering the everyday practices of individuals so as to shape the organization of biosocial life more generally. This includes inculcating the awareness of biological risks, shaping individuals as subjects who relate to themselves as *biological* beings in their everyday activities. The hand sanitizer dispensers discussed at the beginning of this chapter play precisely such a role: beyond their instrumental value for slowing the spread of infectious disease, their most important effect may be to present the world to users or passers-by as seething with biological threats, including the threat posed by one's co-worker or neighbour (whether human or nonhuman). Other examples include vaccination centres, often set up at public sites or in large institutions, instructions to build body immunity, instructions to stay home if feeling sick, to avoid touching eyes, nose or mouth, and to not share such things as towels with others. While none of this advice is new, what is notable is that such practices seek to at once sustain and interrupt flows, and to do so simultaneously. Thus it is not the kind of 'containment' that characterized disciplinary societies (Foucault, 1977), a form of power that is in many respects antithetical to neoliberal economies, but instead approximates much more closely the 'ceaseless control in open sites' that Deleuze (1995) thought characterized societies of control. Indeed, what preparedness planning accepts is precisely that, in a networked world, biological risks cannot simply be 'contained'.

Pre-emption, on the other hand, works on a very different biopolitical register. Unlike preparedness, which seeks to order life in advance of the event so as to mitigate its effects, pre-emption seeks to order life in such a way that the event does not happen at all. Pre-emptive practices are thus substantially different, in that they seek to intervene in the *conditions of emergence*, not unlike counter-insurgency operations that seek to ward off the emergence of 'insurgent' subjectivities (see Anderson, 2010). Pre-emptive power is a form of power that seeks to produce alternative futures, to *determine* that which is determined to be determined in one way rather than another way. Examples range from efforts to transform agricultural practices that are seen as potential sources of biological risks in global networks (such as separating humans and animals and regulating human–animal contact in peasant or subsistence farming), to rapid response teams that can be deployed globally to extinguish outbreaks before they become epidemic, to culling domestic and wild animals (often by the tens of millions) in order to pre-empt the possibility of animal-to-human transmission (see Braun, 2008).

It is with such pre-emptive practices that we can most clearly see why biosecurity can never free itself from its *thanatological* dimensions. Interpreters of Michel Foucault's work on biopower often miss this point. Most correctly understand Foucault's central argument that over the course of the seventeenth and eighteenth centuries *sovereign* power – the power to *take* life or *let* live – was at least partially replaced by forms of governance whose purpose was to *make* live, either by maximizing the power of the body's forces, or by investing in the vitality of populations through such practices as public health, town planning and hygienics. But many miss Foucault's insistence that this was a power not only to 'make live' but to 'let die', and, moreover, to let die so that others might live better. The key point is that even as Foucault identified biopower's objective as the vitality of populations, he insisted that it *necessarily* excluded some individuals, or, perhaps more to the point, included them *through* their exclusion. As he explained in his lecture of 17 March 1976, investing in life paradoxically entailed killing – or at least abandoning or exposing to risk – some populations in the name of the vitality of valued forms of life (Foucault, 2003). For Foucault, this was figured as a cut performed solely within *human* populations, whereby those lives deemed *not* to add to the vitality of the population – the racialized minority, the infirm, the disabled – were ultimately disposable. Giorgio Agamben (2003) would later gloss this in terms of an indistinction between the human and the animal, whereby some humans come to be more or less associated with 'animality', and thus more or less fully 'human'. But in the case of biosecurity, something else and something more is at stake. Indeed, Foucault's account may not need to be merely recounted, but also extended – for not only does biosecurity at times involve a cut that distinguishes between those humans accorded protection and those abandoned (see below), but the same calculus of life and death extends to the entire biological world, not only figuratively as that which justifies the distinction between humans (i.e. as more or less animal), but as a practice that intervenes in life (and *against* life) in order to secure life. This is the case not only because the 'human' and the 'animal' are ultimately indistinct (as Agamben, 2003, notes), but because, as is made evident in the 'ontostories' that pervade biosecurity, the 'human' and the 'animal' are intimately entangled in a networked world of emergent life (as are, indeed, animals with other animals and plants, an insight that informs much of the one health–one world rhetoric that has become recently popular).

Attending to this 'cut' in the fabric of biological life is thus to attend to the way in which forms of life are accorded differential protection or, in some cases, exposed to death. While in biosecurity practices it is human life that is accorded protection (often by killing animals), this often includes privileging certain forms of human life over others, such as in attempts to transform agricultural modes of existence in some parts of the world, in order to achieve 'security' elsewhere. In this sense, biosecurity's biopolitical practices are inherently *geopolitical*, ensuring the protection of valued forms of life in some places (such as in the global North) through ordering biosocial life in other parts of the world (Braun, 2007). What has perhaps received less attention is the calculus of life and death visited on *animal* populations within these practices, such as in the 'culling' of domestic and wild animals in order to pre-empt the irruption of biological threats among humans. At one level this is a great concern to agriculturalists, whose livelihoods are placed at risk. But it also raises essential questions about the *anthropocentrism* of biosecurity practices: the way in which animal life can be sacrificed – and is, by the millions – so that human life can persist.[15] At such moments biosecurity reveals itself as an excessively *violent* affair, as a thanatopolitics that finds its justification in speciesism and hides behind the distinction between the human and the animal. While few would be willing to argue for leaving infected domestic or even wild flocks alone, the issue has not gone unnoticed, most recently in the case of a shadowy group in Canada called the Farmer's Peace Corps, which recently kidnapped a flock of sheep destined to be slaughtered in accord with biosecurity measures, refusing to identify their location to authorities.[16]

Such actions may be derided for simply increasing biological risks (in this case biological risks that are primarily economic risks), but the actions of the group point to perplexing ethical questions that pervade biosecurities of all sorts and which cannot easily be evaded: questions that range from which life is valued and which is sacrificed, who decides and on what basis, all the way to questions concerning the ways in which food is today produced and industrial agriculture organized (on the latter, see Wallace and Kock, 2012). Behind all this lies perhaps a larger question: whether it might be possible to imagine biosocial assemblages that allow for the flourishing of life (human and nonhuman) beyond the current integration of life and law.

Notes

1　Note the prevalence today of 'one world-one health' rhetoric, which at once explicitly recognizes the entanglement of humans and animals, and seeks to imagine ways to intervene in it.

2　We might add here the call to 'democratize' biosecurity decisions. Doing so does not necessarily escape its biopolitical dimensions, but instead adds more voices to the decision of where such 'cuts' in human-animal networks should be made.

3　For discussion of the range and politics of everyday biosecurity practices, see work associated with the Biosecurity Borderlands Project (www.biosecurity-borderlands. org).

4　It would be incorrect to suggest that there is only one set of ontological presuppositions; what follows is influenced by the examples chosen, but might be said to be generalizable across numerous biosecurity practices.

5　Biosecurity involves 'framing' life in particular ways; in other words, it works with a set of implicit or explicit ontological presuppositions.

6 Arguably biosecurity in North America and Europe emerged within three sites: infectious diseases, food-borne illnesses, and the spectre of bioterrorism.

7 I borrow the phrase 'proliferating life' from Paul Jackson (2010).

8 See Myra Hird (2009) for a discussion of the centrality of bacteria to human (social) life.

9 The turn to a 'more-than-human' urban geography, or what has elsewhere been called 'urban political ecology' was spurred in part through an awareness of cities as *assemblages* of humans and nonhumans (see Braun, 2005; Heynen et al., 2006; Hinchliffe and Whatmore, 2006).

10 The significance of transportation networks was again reinforced during the H1N1 pandemic in 2009, where the virus spread rapidly from Mexico to the United States via the numerous air links between the two countries. The 2011 movie *Contagion* plays on precisely this aspect of global assemblages.

11 A similar argument is put forward by Dillon and Reid (2009), who understand security as taking as its concern 'bodies in formation', wherein bodies (the infectious body, the body of the terrorist, the insurgent body) are seen as emergent effects of global networks characterized by 'circulation, connectivity and contingency'.

12 Archival-statistical knowledge allowed for forms of calculative rationality traditionally associated with insurance (for a detailed discussion, see Collier, 2008).

13 Collier (2008) describes such methods as 'enactment based'.

14 This has included attempts to reconstruct the virus in order to study it, something that has carried with it its own risks and security practices.

15 During the HPAI scare of 2004, more than 62 million birds in Thailand were slaughtered. See wwwnc.cdc.gov/eid/article/11/11/05-0608_article.htm.

16 See www.theglobeandmail.com/news/national/ontario-sheep-kidnappers-say-infected-flock-is-in-protective-custody/article2417448.

References

Agamben, G. (2003) *The open: man and animal*, Stanford University Press, Stanford, CA.

Anderson, B. (2010) 'Preemption, precaution, preparedness: anticipatory action and future geographies', *Progress in Human Geography*, vol. 34, pp. 777–798.

Barker, K. (2010) 'Biosecure citizenship: politicizing symbiotic associations and the construction of biological threat', *Transactions of the Institute of British Geographers*, vol. 35, pp. 350–363.

Bennett, J. (2010) *Vibrant matter: a political ecology of things*, Duke University Press, Durham, NC.

Braun, B. (2005) 'Writing a more-than-human urban geography', *Progress in Human Geography*, vol. 29, no. 5, pp. 635–650.

—— (2007) 'Biopolitics and the molecularization of life', *Cultural Geographies*, vol. 14, pp. 6–18.

—— (2008) 'Thinking the city through SARS: bodies, topologies, politics', in H. Ali and R. Keil (eds) *Networked disease: emerging infections in the global city*, Blackwell, Oxford, pp. 250–266.

CIDRAP (2007) 'The pandemic vaccine puzzle', Center for Infectious Disease Research and Policy, www.cidrap.umn.edu/cidrap/content/influenza/panflu/news/oct2507pan vax1.html, accessed 25 May 2012.

Collier, S. (2008) 'Enacting catastrophe: preparedness, insurance, budgetary rationalization', *Economy and Society*, vol. 37, no. 2, pp. 224–250.

Cooper, M. (2008) *Life as surplus*, University of Washington Press, Seattle, WA.

Craddock, S. (2004) *City of plagues: disease, poverty and deviance in San Francisco*, University of Minnesota Press, Minneapolis.

Davis, M. (2005) *The monster at our door: the global threat of avian flu*, New Press, New York.

Deleuze, G. (1995) 'Postscript on societies of control', in G. Deleuze (ed.) *Negotiations 1972–1990*, Columbia University Press, New York, pp. 177–182.

Dillon, M. and Reid, J. (2009) *The liberal way of war: killing to make life live*, Routledge, London.

Foucault, M. (1977) *Discipline and punish*, Penguin, London.

——— (1991) 'Governmentality', in G. Burchell, C. Gordon and P. Miller (eds) *The Foucault effect: studies in governmentality*, University of Chicago Press, Chicago, pp. 87–104.

——— (2003) *Society must be defended: lectures at the Collège de France 1975–76*, Penguin, London.

Gandy, M. (2006) 'The bacteriological city and its discontents', *Historical Geography*, vol. 34, pp. 14–25.

Halloran, M. E., Ferguson, N. M., Eubank, S., Longini, I. M., Cummings, D. A. T., Lewis, B., Xu, S., Fraser, C., Vullikanti, A., Germann, T. C., Wagener, D., Beckman, R., Kadau, K., Barrett, C., Macken, C. A., Burke, D. S. and Cooley, P. (2008) 'Modeling targeted layered containment of an influenza pandemic in the United States', *Proceedings of the National Academy of Sciences of the USA*, vol. 105, no. 12, pp. 4639–4644.

Heynen, N., Kaika, M. and Swyngedouw, E. (eds) (2006) *In the nature of cities: urban political ecology and the politics of urban metabolism*, Routledge, London.

Hinchliffe, S. and Whatmore, S. (2006) 'Living cities: towards a politics of conviviality', *Science as Culture*, vol. 15, pp. 123–138.

Hird, M. (2009) *The origins of sociable life: evolution after science studies*, Palgrave Macmillan, Basingstoke, UK.

Jackson, P. (2010) 'Cholera and crisis: state health and the geography of future epidemics', Ph.D. Dissertation, University of Toronto.

Massumi, B. (2009) 'National enterprise emergency: steps toward an ecology of powers', *Theory, Culture and Society*, vol. 26, no. 6, pp. 153–186.

Mitchell, T. (2002) *Rule of experts: Egypt, techno-politics, modernity*, University of California Press, Berkeley.

Reid, J. (2007) *The biopolitics of the war on terror: life struggles, liberal modernity and the defence of logistical societies*, Manchester University Press, Manchester.

Smith, R. G. (2003) 'World city topologies', *Progress in Human Geography*, vol. 27, no. 5, pp. 561–582.

Sparke, M. and Anguelov, D. (2011) 'H1N1, globalization and the epidemiology of inequality', *Health and Place*, vol. 18, pp. 726–736.

Spinoza, B. (1996) [1677] *Ethics*, Translated by Edwin Curley, Penguin, London.

Thacker, E. (2005) *The global genome: biotechnology, politics, and culture*, MIT Press, Cambridge, MA.

Wallace, R. G. and Kock, R. A. (2012) 'Whose food footprint? Capitalism, agriculture and the environment', *Human Geography*, vol. 5, no. 1, pp. 63–83.

Webster, R. (2003) 'The world is teetering on the edge of a pandemic that could kill a large fraction of the human population', *American Scientist*, vol. 91, no. 2, p. 122.

Part II

IMPLEMENTING BIOSECURITY

4

GOVERNING BIOSECURITY

Andrew Donaldson

Introduction

This chapter is concerned with the different frameworks for governing biosecurity that currently exist. The United Nations Food and Agriculture Organization (FAO) offers a definition of biosecurity as being 'the process and objective of managing biological risks' to environment, food and agriculture in a holistic manner (FAO, 2003, p. 1). This definition is an attempt to bring some clarity to a term that has been used with increasing frequency in policy circles, while still keeping a breadth of meaning that can incorporate various understandings. It is also part of an attempt by the FAO to use the prominence and novelty of biosecurity to drive policy makers of national governments in a particular direction.

Although biosecurity policy and regulation takes place on an international stage, there is no article of international law in which the term biosecurity appears (Manzella and Vapnek, 2007). While the FAO has a guiding role, and is arguably the only international organization promoting biosecurity as an integrative term, there are several international bodies with a direct material influence over the shape and substance of national frameworks for producing biosecurity. It is in response to the work of these organizations that the FAO has adopted the concept of biosecurity and dedicated a specific work programme to it. The OIE (Office International des Epizooties, now usually translated as the World Organization for Animal Health) is the oldest of these organizations, having been set up in 1924 in response to a devastating outbreak of rinderpest in Europe, caused by zebu oxen (*Bos primigenius indicus*) being transported from India to Brazil via Antwerp. The OIE produces standards for the animal health services of its 172 members including manuals of specified diagnostics and vaccines. The OIE Codes are key international standards for determining reasonable measures that may be taken by a state to protect animal health. They have been given extra potency by the adoption of the OIE as a standards reference body for the World Trade Organization's (WTO) Sanitary and Phytosanitary (SPS) agreement. As long as the states concerned follow the OIE Codes it is unlikely that any import or export bans would be deemed unfair barriers to trade by the WTO. Under the SPS agreement, the International

Plant Protection Convention (hosted by the FAO itself) and the Codex Alimentarius Commission play the same roles with respect to plant health and food safety. The Convention on Biological Diversity, and its associated Cartagena Protocol on Biosafety, has a particular concern with problems of regulating and managing invasive species and is a further influence on the FAO's work.

Some international institutions, agreements and codes of practice have been in existence for longer than the term 'biosecurity' itself, with some national systems pre-dating the international institutions. Biosecurity is a relatively novel term that has risen up the international agenda (Donaldson, 2008; FAO, 2008; Falk et al., 2011). In accounting for this, the FAO characterizes biosecurity as a new strategic viewpoint from which governments can consider existing arrangements for plant, animal, human and environmental health in light of an increasingly interconnected planet. According to the FAO (2003, 2007, 2008) biosecurity frameworks can be classified into two types: traditional (or sectoral) and integrated. The traditional approach is predicated on separate consideration of the various sectors (animal health, plant health, food safety, environmental protection, etc.) in which policy and regulation are likely to have developed separately over time, resulting in a fragmented landscape with contradictory laws, inefficient use of resources and gaps in research. The integrated approach is that promoted by the FAO through its Biosecurity Toolkit and focuses on a strategic deployment of resources through measures such as cross-sector collaboration, harmonization of legal frameworks and joint priority setting. This distinction serves as a starting point for this chapter, which draws on comparison between biosecurity arrangements in the UK and New Zealand, adding examples from other countries where this is useful. While the UK embodies the traditional or sectoral approach to biosecurity, with no primary legislation that uses the term biosecurity, New Zealand is usually taken as the longest established example of an integrated approach. I detail the differences and similarities between the two and develop further the discussion of integrated biosecurity frameworks, globally. This is followed by an examination of the way in which risk analysis dominates biosecurity and the ways in which governments are now attempting to share biosecurity risks more formally with industry. I briefly consider the way in which groups beyond the agrifood sector figure in state biosecurity frameworks.

Running through all of this are questions around the politics of biosecurity. No system of biosecurity governance currently in existence considers the issues it deals with to be 'political'. That is, they are not matters that should be open to ongoing debate and questioning; they are self-evident problems to be resolved through technical means. In this vein, the FAO promotes biosecurity as a strategic policy device for rationalizing government. But biosecurity can also be characterized as a huge socio-technical experiment (Latour, 2001) occurring on multiple scales, involving a range of different regulatory systems and public and private sector actors. The questions this approach prompts are not about the best way to organize biosecurity policy, but about what should be done in the face of uncertainty, whose knowledge should count and, ultimately, what is actually at stake when we talk about biosecurity.

Policy and regulatory frameworks

The development of the policy and regulatory framework for biosecurity in the UK has emerged from a long and disjointed history of interventions in agriculture, principally around animal health. Indeed, animal health became an object of government before agriculture in general through a series of political moves that put the veterinary profession at the heart of government policy (Enticott et al., 2011). As a series of high-profile outbreaks in recent decades has shown – from bovine spongiform encephalopathy (BSE) to foot-and-mouth disease (FMD) and the ongoing problem of bovine tuberculosis – animal diseases have far-reaching political, economic and public health consequences. Animal health policy and regulation sits within an international and European framework. At the international level, the UK is a member of the OIE along with other EU member states. The EU as a whole also aims to meet the requirements of the OIE, as supported by the WTO, in order to maintain its common market and operate as a global trade bloc. These standards are implemented in EU legislation governing veterinary checks for trade in animals and animal products and in the specified control measures to be taken against certain animal diseases, including preparation of a National Control Plan with information on the structure, relationships and responsibilities of the various 'competent authorities' for implementing animal health and food safety legislation. The main piece of legislation governing animal health in England is the Animal Health Act 1981, as amended (most notably in 2002, following the 2001 foot-and-mouth disease epidemic). There are around 175 statutory instruments applying to animal health issues, with most made under the Animal Health Act. The Animal Health Act permits for infringements of secondary legislation made under it to be considered offences and prosecuted.

In England alone, the bodies responsible for developing and delivering policy around these various sets of regulations are myriad. They include one central government department (Defra – the Department for Environment, Food and Rural Affairs) and six of its executive agencies (including Animal Health); two non-ministerial departments (the Food Standards Agency, which absorbed the Meat Hygiene Service executive agency responsible for inspection at abattoirs in 2010, and HM Revenue and Customs); 149 local authorities and their representative body, the Local Authorities Coordinator for Regulatory Services (LACORS); Port Health Authorities; and a large number of veterinarians in private practice who also undertake government work as Official Veterinary Surgeons.

Plant health in the UK has lagged behind animal health as an interest of state, possibly through lack of an equivalent to the powerful livestock and veterinary lobby that has driven state intervention in animal health (Waage and Mumford, 2008). Key legislation here is the 1967 Plant Health Act (amended by the Plant

Health Order 2005, to bring it into line with EU legislation) and in excess of 100 statutory instruments. Implementation is more straightforward and centralized than for animal health, being led by Fera (the Food and Environment Research Agency – a former agency of Defra, now fully privatized) along with the Forestry Commission and relevant departments of the Scottish Government and Northern Ireland assembly.

This brief overview of the UK framework serves to highlight the complexity of the field. There are moves towards greater integration within sectors (for example, the RADAR system for collecting animal health and veterinary data together) and there is some collaboration across sectors, with the FSA and its responsibility for food safety bridging the gap between animal health and plant health and working with bodies on both sides. Fera, responsible for plant health, also has significant expertise in a range of environmental issues and houses the Bee Health unit. With regard to border control, all relevant bodies for plant and animal health and food safety meet twice yearly with HMRC and the UK Border Agency for coordination purposes. A single overarching document that highlights the ways in which the various bodies work together to meet food and feed laws, animal and plant health, has existed since 2007 in the form of the Single Integrated National Control Plan for the United Kingdom or National Control Plan for short. This document covers a 5-year programming period and was produced to meet EU requirements. The FSA takes the lead on preparation and maintenance of the plan. Despite this outward nod towards integration, the idea of biosecurity as an overarching concept is absent from UK policy. The term is still found mainly in respect to hygiene practices at livestock premises, although an obvious exception to this is the name of the recent Tree Health and Plant Biosecurity Action Plan (Defra and Forestry Commission, 2011). The haphazard way in which the term is used is highlighted by comparison with the equivalent document for the animal health sector, the Animal Health and Welfare Strategy (Defra, 2004).

In contrast to the sectoral approach of the UK, New Zealand has adopted the integrated approach. The idea of biosecurity was part of national policy and law (via the 1993 Biosecurity Act) in New Zealand long before the term was even in use in the UK (Donaldson and Wood, 2004). The existence of primary legislation concerning biosecurity, along with a Minister for Biosecurity post, immediately sets New Zealand apart from the UK. In 2003, New Zealand's first biosecurity strategy was introduced that led to the formation of Biosecurity New Zealand, a new agency which consolidated a range of central government biosecurity functions and formed a point of liaison with industry and other key government bodies, such as the Department of Conservation, Ministry of Health and the Environmental Risk Management Authority. In 2007, the agency merged with the border control functions of its parent Ministry of Agriculture and Forestry (MAF) to form MAF Biosecurity New Zealand (MAFBNZ), which provides 'leadership' for biosecurity in New Zealand. MAFBNZ maintains a website dedicated to bringing all biosecurity-related policy and information together; the agency represents a significant strengthening of the institutional capacity for cross-sector collaboration that brings the organizational framework further into line with the integrated approach signalled by the 1993 Act. As an outward face of biosecurity in New Zealand, this indicates a clear distinction between the UK's traditional approach and the idea of an integrated approach to biosecurity. The next clear difference is in the characterization of the various policy and delivery bodies, including industry bodies and third sector environmental groups, as constituting a 'biosecurity system'. This system is conceptualized as operating in a series of overlapping geographical 'zones of activity': global, pathways and borders, and within New Zealand. The majority of the work undertaken within these zones is orientated around the maintenance of borders and being able to safely import and export goods. Activities in the global sphere include both disease surveillance and involvement in international biosecurity policy development. Under pathways and borders, the biosecurity system is concerned with the management of risk before threats enter the country and at the border, a process of monitoring and licensing as well as stringent border regulation (including the famous airport sniffer dogs that search for illegally imported biological materials). Within New Zealand, the emphasis is on the management of already existing 'pests' – the catch-all term used in the 1993 Act for undesired organisms. Even when considering diseases, the emphasis is on strict border controls as the best response.

When considered alongside the UK and European context, New Zealand offers a considerably more focused and coherent approach to biosecurity. However, it is not as comprehensive an approach as put forward through EU requirements. The National Control Plans that must be prepared by EU member states, regardless of their national systems, must include elements that are downplayed, at least as biosecurity, in New Zealand. There, food safety is notably set to one side of biosecurity, with the two always mentioned separately in the Statement of Intent (MAF, 2011) – the document setting out strategic activities of the ministry – to the point that the food system and the biosecurity system are given distinct mentions. On this evidence, integrated biosecurity has the potential to create policy silos just as the sectoral approach is assumed to.

There is some evidence that the integrated approach is being promoted as a desirable global norm. The influence of the FAO on agricultural trade in the Global South provides a ready channel for their vision to be translated into emerging systems, with the integrated approach being sold as an efficient measure to help build agrifood legislation from scratch or to reform overly complicated legislation (see Manzella and Vapnek, 2007). The 'offshore' biosecurity work of countries that have adopted the integrated model for their own systems is another channel by which this approach is spread. A key example documented here is Australia, which adopted an integrated model of biosecurity akin to New Zealand's approach through the replacement of 17 separate pieces of agricultural legislation with the Biosecurity and Agricultural Management Act 2007. Australian public and scientific bodies are now heavily involved in helping to develop better biosecurity within Indonesia, a country in which half the population lacks basic clean water and sanitation measures, and which is viewed as a major biosecurity threat to Australia, owing to its relative proximity (Falk et al., 2011).

While the FAO may promote biosecurity as an integrative, strategic concept, its position within state institutions is variable. Biosecurity can entail very different ways of assembling a variety of components. Biosecurity in the UK is subsumed by diverse animal, plant and public health policies. Conversely, biosecurity in New Zealand incorporates concerns about animal and plant health environmental risk. These different arrangements are fundamentally experimental framings of a series of complex and uncertain problems that manifest in different ways over time and space. No system of governing biosecurity is provably better than another with respect to the goal of preventing pathogen or pest incursions. There are also indications that the term has political resonances beyond the practicalities of public administration. The adoption of an integrated approach by Australia, which is then inculcated in Indonesia, suggests that creating integrated biosecurity policy, in line with the FAO's recommendations, might be a way in which emerging economies signal their worthiness as trade partners to those in the driving seat of biosecurity and trade policy.

Surveillance and risk

Regardless of whether they follow a sectoral or integrated model, all biosecurity frameworks rely on forms of surveillance, whether these be for unwanted alien species, animal or plant diseases, or human health threats. Evaluating, classifying and monitoring the various pathways through which biosecurity threats might manifest is long-established and fundamental to all existing biosecurity systems, appearing at all scales of intervention (Donaldson and Wood, 2004). While surveillance in general has a very public politics centred on issues of privacy, within a biosecurity context it often goes unnoticed or unremarked upon outside of agribusiness or farming communities, where concerns may be raised over the perceived intrusion of state monitoring. As more people are drawn into the surveillant networks of biosecurity (see below), we might expect increased questioning of why this monitoring is necessary: What are we actually working to protect? Currently, surveillance in the governing of biosecurity is effectively depoliticized through recourse to risk analysis.

Biosecurity is fundamentally about dealing with risk; that is, dealing with problems before they are actualized. When dealing with threats to human, animal and plant health the motif of 'prevention is better than cure' dominates. This involves the identification of pathways and vectors by which biosecurity threats might arrive in a country, the likelihood of such events occurring and their potential impacts. The FAO recognized risk analysis as a 'common framework' for biosecurity in its original agenda-setting session on the topic (FAO, 2003), because risk analysis was already a 'common thread among the many international instruments relevant to biosecurity' (Manzella and Vapnek, 2007, p. 8). Key among these is the WTO SPS agreement, which requires members to establish an 'appropriate level of protection' based on either the international standard or a risk analysis approach. For animal and plant diseases and invasive pests, the pathways for entry and spread must be evaluated along with potential impacts as noted above; in the case of food safety issues only the adverse economic and biological effects need to be considered. The SPS agreement also permits an implementation of the precautionary principle, by making provision for protective measures to be taken in the absence of a standard or full risk assessment when there is insufficient knowledge to undertake the assessment, provided that knowledge is then sought. While the various organizations referenced by the SPS agreement all utilize forms of risk analysis independently (as does the Convention on Biological Diversity (CBD), which falls outside of the agreement), it is their bringing together under a common risk-based framework that provides an international basis for integrated biosecurity approaches (Manzella and Vapnek, 2007). As following the tenets of the SPS agreement is a basis for global trade, there is significant pressure for countries to adopt or enhance risk-based approaches in their own biosecurity frameworks.

Risk analysis is problematic. Apart from requiring up-to-date data (via constant surveillance) and continuous research into emerging threats, the actual outcomes can be difficult to interpret and integrate. This is particularly the case with the SPS agreement's 'appropriate level of protection', which is also referred to as the 'acceptable level of risk', which can be difficult to determine (Mumford, 2002). While risk analysis forms a high-level common framework, the tools and approaches often vary widely in different sectors. Manzella and Vapnek (2007) note that consolidation of risk analysis tools across sectors is a key component of an integrated biosecurity framework as proposed by the FAO; as well as a technically useful solution this integration is promoted as an economic efficiency by the FAO. Waage and Mumford (2008) have argued for an extension and consolidation of biosecurity risk analysis in the UK, bringing together animal and plant measures and also developing a more sophisticated approach to considering risks that focuses on the desired outcome rather than the action to consider different routes of reaching it. The example they use is developing disease-resistant crops rather than implementing tighter border controls – both ways of managing the risk of new plant disease. They also note that an important driver for improving risk analysis is the ability to target stretched resources better.

Risk analysis is far from an objective technical practice for a number of reasons. It is entangled in international trade politics, effectively supplying a negotiable argument for biosecurity decisions, rather than a definitive standard. Outcomes that are difficult to interpret raise the likelihood of competing knowledge claims. Deciding on 'acceptable' levels of risk begs the question: acceptable to whom? Waage and Mumford's (2008) suggestion of focusing on desired outcomes raises the parallel question: desired by whom? These are fundamentally not 'expert' matters; they are questions concerning the public good and can be put to a wider constituency. However, risk thinking has driven the involvement of wider publics in a different direction.

Managing risk is not purely about prevention; a risk analysis must take into account the exposure and vulnerability of that which is under threat. So a risk-based approach must also consider what will come to pass if a risk is actualized and the ramifications of dealing with a disease event or similar. This entails contingency planning, which in the EU is now a legal requirement for certain diseases, and was a major preoccupation in the UK following the 2001 foot-and-mouth disease epizootic. The routine business of biosecurity is uneventful for many of those involved. Absence of a problem is a defining feature of a functioning system. In the event of a biosecurity breach, different sets of powers come into play and the calculation of risk takes on a greater urgency and a finer granularity. In Defra in the UK, this switchover is referred to as moving from 'peace time' to 'war time', reflecting the fact that the department, and its predecessor ministries, have been built around dealing with disease events (Wilkinson, 2011). But having to deal with problems that break out in such spectacular fashion also opens up new risks for biosecurity bodies. Biosecurity risks can be classed as 'societal risks'; and dealing with societal risks opens up new 'institutional risks' (Rothstein et al., 2006; Donaldson, 2008). These institutional risks are the reputational, operational and financial costs that an organization that 'owns' a particular societal risk must face in managing that societal risk. (In the risk registers that governments and other organizations produce, risks are said to be owned by the actor with lead responsibility for managing the risk.) The management of secondary institutional risks may be a factor that has lent urgency to recent attempts to share responsibility for dealing with primary biosecurity risks.

Paying for biosecurity

One common thread driving the rise of biosecurity programmes around the world is that of increasing biological risk. Climate change will have increasing effects on the ranges of species, leading to increased incursions by invasive species and the wider spread of disease vectors – for example, the midge that spreads bluetongue virus among ruminants has already spread through northern Europe as a result of climate change (Jones, 2011). In the UK, successive governments have expressed mounting concern about the cost of biosecurity, prompted by the scale of recent exotic disease problems and escalating endemic disease problems. The 2001 FMD epizootic in the UK cost around £8 billion and had significant impacts on both the public purse and on industry outside of agriculture (see Donaldson et al., 2006); similarly a large increase in the incidence of endemic bovine tuberculosis had led to huge increases in the budget required to manage the disease. These concerns led to the introduction of a responsibility- and cost-sharing (RCS) agenda in the UK with the 2004 Animal Health and Welfare Strategy (Defra 2004). At that time the body responsible for driving forward the overall strategy oversaw the development of the cost-sharing agenda. In 2006, a new stakeholder group, the UK Responsibility and Cost Sharing Consultative Forum, was set up to make recommendations on a partnership approach for joint development, delivery and funding of animal health policy. The Forum was wound up in 2008, and in 2009 a new Advisory Group on Responsibility and Cost Sharing was set up to provide further advice. In January 2010 Defra put a draft Animal Health Bill forward for consultation. The bill would have put into law provisions for cost sharing and created a new Animal Health Organization, which would have been able to raise a levy from industry for disease control and biosecurity operations. The bill also made provisions for reducing compensation payment to persons who had contributed to the spread of disease.

This form of hard regulation was mooted as a means of changing farmer behaviour. A voluntary approach to encouraging responsibility sharing had been attempted with promotion of farm health planning. This was intended to get livestock keepers to take a more planned and active responsibility for the health of their animals; it was couched in terms of being good business sense, and the incentive for creating a farm health plan was set out by the slogan, 'healthy animals, healthy profits'. Farm health planning gradually slid down the agenda, with farmers generally either already engaged in industry standard good practices (in the more lucrative intensive pig and poultry sectors) or showing little interest. Early attempts to link farm health planning and good on-farm biosecurity practices to the payment of compensation for livestock losses in the event of a disease outbreak had been abandoned owing to the difficulty of determining what would constitute good biosecurity in a measurable way (Donaldson, 2008). The European Commission has taken a stance that compensation payments themselves might be an active disincentive to disease prevention and that cost sharing would promote greater biosecurity (Anonymous, 2006).

While the RCS Advisory Group survived the cull of such bodies by the incoming coalition UK government in 2010, the Animal Health Bill itself was dropped. The new government did, however, remain committed to the RCS agenda and, following the final recommendations of the Advisory Group, set up in 2011 the Animal Health and Welfare Board for England. The new body will review policy and make recommendations about charging as it sees appropriate. At the time of writing it is almost 8 years since the RCS agenda was formally introduced in the UK. Following all of the changes in governance and the multiple rounds of consultation and advice outlined here, there is still no definitive answer to the question of whether or how to share the costs of UK biosecurity between the public and private sectors, and opposition from the livestock producers' lobby remains strong. An argument in favour of cost sharing that is rising to greater prominence is the need for greater fairness in the treatment of animal and plant health issues (Waage and Mumford, 2008; Wilkinson et al., 2010). As noted in the previous section, the agricultural state in the UK is heavily influenced by a powerful livestock and veterinary lobby, and plant health issues have never received the same financial support from the state as animal health. Producers in the plant-based sectors have borne all the costs of biosecurity and disease outbreaks themselves.

Cost sharing for biosecurity in New Zealand has taken on a slightly different form. It was introduced in 2009 – following 4 years of consultation through a government–industry joint venture by the Surveillance and Incursion Response Working Group – as part of a new package of biosecurity measures aimed at strengthening a partnership approach with industry. Here, cost sharing was paired with joint decision making, rather than a rhetoric of sharing responsibility. In essence this joint decision making can be regarded as responsibility sharing, but with a different angle than that taken in the UK. The difference in rhetoric may also be due to a generally different view of responsibility with regard to biosecurity than can be found in New Zealand (see next section). The instrument for establishing joint decision making and cost sharing is the Government Industry Agreement, presented with the slogan, 'Building a biosecurity system is a collaborative project. It takes a whole country' (MAFBNZ, 2009). Cost sharing under the agreement is to be via a levy on industry. Against claims that it was attempting to pass the costs of biosecurity onto producers, MAF asserts that it is not reducing its spending on biosecurity and that the additional spending from industry will effectively buy a greater say in how biosecurity resources as a whole are deployed (the joint decision-making component). This is in direct opposition to the position in the UK, where Defra is clear that cost sharing is needed in order to reduce state spending. Another contrast to the UK is that there is already greater equality in the treatment of plant and animal health, with the 1993 Biosecurity Act not discriminating between the two in terms of outcomes.

Resistance to cost sharing in the UK may not be without sound foundation. The livestock sector has been used to a top-down regime based on regulation and financial compensation from government, so proposed responsibility and cost sharing represent a 'culture shock', but there are other reasons for concern (Wilkinson et al., 2010). First, disease can be spread in ways that farmers can do little about (airborne or via difficult-to-detect vectors); second, the payment of compensation is a means of encouraging them to report incidence of disease (Wilkinson et al., 2010). In contrast to UK producers' responses, the objections raised by the Federated Farmers of New Zealand (FFNZ) to measures proposed there were based on their fiscal appropriateness (FFNZ, 2009). Their argument was that biosecurity was a matter of national security that was paid for through normal taxation and, given that farmers had no influence on what happened at borders, they should not have to pay again for biosecurity. This is, of course, of a piece with the way in which biosecurity has been developed and promoted as an integrated feature of national security (especially with respect to defending the natural environment against alien species) in New Zealand.

The public-good nature of biosecurity is often put at the centre of arguments for and against cost sharing. In New Zealand, MAF has recently published an outline of the economic theory it considers important to deciding on cost-sharing mechanisms, claiming that biosecurity exhibits both public good and club good characteristics (MAFBNZ, 2011); consumption of the good does not deplete it, but some aspects of it are excludable. For example, public health benefits are shared by all, but the profits accruing from the ability to trade freely are not. The report argues that willingness to pay is a good indicator of the viability of framing a particular benefit as a club good and so provides a number of classes of biosecurity benefits for which the state and industry will be willing or unwilling to pay. There are significant definitional problems in these debates, as any given actors' willingness to accept the economic arguments very much depends on their understanding of biosecurity (what it entails in practice and the extent to which it actually works); their view of who benefits (marginal producers may be unwilling to invest in programmes that will disproportionately benefit larger operations); and indeed their notion of a public good (this may be an objective matter for some, a normative or value-driven concept for others) or the role of the state (the level of intervention that is acceptable in the name of security).

Engaging beyond producers

Cost sharing, in the instances described above, is an attempt by governments to find a new way of working with producers to maintain biosecurity. What means exist of engaging those players with a less direct financial interest in biosecurity?

In some cases there may be a useful alignment of the goals of certain publics and biosecurity outcomes, potentially encapsulated through third sector representation. A key example here is the Association of River Trusts in the UK, which brings together various stakeholder bodies concerned with the quality of rivers. Anglers want high-quality rivers that provide them with better sport fishing opportunities. Through the River Trusts they can engage with catchment biosecurity plans and provide additional surveillance for disease and invasive species in their localities (Owen, 2011).

As previously highlighted, the construction and communication of biosecurity policy in New Zealand is framed in terms of biosecurity being a national responsibility that 'takes a whole country'. Barker (2010) has investigated this 'biosecure citizenship' and documented the ways in which the inhabitants of New Zealand both have a formal duty to report suspected invasive organisms and are also encouraged to voluntarily engage in other management activities. Barker questions whether this mode of public engagement is appropriate given the lack of involvement the citizens have in determining the goals of biosecurity practice.

In Australia, research has been conducted on how better to involve individuals in achieving biosecurity goals. The 'Engaging in Biosecurity' project (Kruger et al., 2009) investigated how to enrol the wider public in surveillance activities. The researchers suggest that it is important to manage media messages and focus on building trusted relationships between biosecurity agencies and the public. This can be accomplished by going beyond traditional approaches to natural resource management in which state agents set out to teach communities about the correct course of action. It is argued that a two-way communication between local communities and biosecurity authorities helps to build a sense of partnership. In direct comparison to the New Zealand example outlined above, the emphasis on wider public involvement does not consider involvement in agenda setting or goal setting. Here, the building of trusting relationships is a means by which government can better target biosecurity messages (and be assured of their favourable reception) and improve the monitoring and reporting of threats by local communities. It is a further enrolment of citizens into the surveillance apparatus.

There is a sense here that biosecurity (in a manner associated with surveillance in general) is 'creeping' into ever more areas of life. If people are to be asked to participate in biosecurity practices, they ought also to have a say in the agenda setting behind those practices. Otherwise, there is a legitimate argument that they are merely working in the interest of powerful lobbies that stand to benefit from biosecurity regulation.

Conclusion

The distinction between integrated and sectoral approaches to biosecurity that I have used as a framing device is not a matter of 'which is better'; it is interesting because of the way it illuminates the politics of biosecurity. First, there are the powerful interests that have driven the uneven development of the sectoral approach, exemplified by animal health in the UK. Second, there is the power of the term biosecurity itself, embodied in the integrated approach that is promoted as a way for countries to become good global citizens and safe trading partners. The commonalities found across different frameworks are also interesting. Cost sharing and public involvement have been mooted or implemented widely, giving rise to debate among various constituencies, as noted above. Surveillance and risk analysis form the basis of all biosecurity governance and have their own less obvious politics.

While biosecurity is fundamentally concerned with risk, I have written elsewhere (Donaldson, 2008) that there are different ways to respond to risk: calculate and monitor, or act experimentally and continually question. No system of biosecurity governance currently in existence has an inbuilt means for questioning the assumptions on which it is built or for allowing such questioning to occur; this is rather subject to broader sets of state priorities and thus to the problems of participation and agenda setting common to political debate and policy formulation. Wider public participation in biosecurity when it happens leans heavily on a logic of risk management (enrolling citizens into biosecurity's surveillance practices), rather than one of accountability and challenge. With biosecurity, the impossibility, and undesirability, of total surveillance and the inherent indeterminacy of risk, along with increasing ecological uncertainty, mean that we are always taking a risk. Yet the range of options from which to select is currently restricted by long established institutional frameworks that channel biosecurity in certain directions. What was a means of protecting health and trade has itself become the sole end of governing biosecurity. The experimental question, What is at stake when we consider biosecurity?, is nothing less than how we live in the world and manage our relationships with other species. Waage and Mumford (2008), in proposing a form of risk analysis that is more flexible and inclusive of different options, do not couch their argument as political, and yet the best way to expand the range of options for biosecurity interventions might be to open up the process to more diverse publics. The government of biosecurity needs to become more creative in how the problems it contends with are imagined.

References

Anonymous (2006) 'Cost sharing in Europe', *Veterinary Record*, vol. 159, no. 11, p. 329.

Barker, K. (2010) 'Biosecure citizenship: politicising symbiotic associations and the construction of biological threat', *Transactions of the Institute of British Geographers*, vol. 35, pp. 350–63.

Defra (2004) *Animal Health and Welfare Strategy*, Department for Environment, Food and Rural Affairs, London.

Defra and Forestry Commission (2011) *Action Plan for Tree Health and Plant Biosecurity*, Department for Environment, Food and Rural Affairs, London.

Donaldson, A. (2008) 'Biosecurity after the event: risk politics and animal disease', *Environment and Planning A*, vol. 40, no. 7, pp. 1552–67.

Donaldson, A. and Wood, D. (2004) 'Surveilling strange materialities: categorisation in the evolving geographies of FMD biosecurity', *Environment and Planning D: Society and Space*, vol. 22, no. 3, pp. 373–91.

Donaldson, A., Lee, R., Ward, N. and Wilkinson, K. (2006) 'Foot and mouth – five years on: the legacy of the 2001 foot and mouth crisis for farming and the British countryside', Newcastle University, Centre for Rural Economy Discussion Paper Series, no. 6, www.ncl.ac.uk/cre/publish/discussionpapers/pdfs/dp6.pdf.

Enticott, G., Donaldson, A., Lowe, P., Power, M., Proctor, P. and Wilkinson, K. (2011) 'The changing role of veterinary expertise in the food chain', *Philosophical Transactions of the Royal Society B*, vol. 366 no. 1573, pp. 1955–65.

Falk, I., Wallace, R. and Ndoen, M. L. (eds) (2011) *Managing Biosecurity Across Borders*, Springer, London.

FAO (2003) *Biosecurity in Food and Agriculture (Committee on Agriculture, Seventeenth Session)*, United Nations Food and Agriculture Organization, Rome.

—— (2007) *Biosecurity Toolkit*, United Nations Food and Agriculture Organization, Rome.

—— (2008) 'Capacity building for standards compliance and certification', BAFRA National SPS Workshops, 24 December.

FFNZ (2009) 'Government gives biosecurity hospital pass', Media release, 3 September, www.fedfarm.org.nz/n1663.html.

Jones, T. (2011) 'Bluetongue outbreaks set to rise with climate change', *Planet Earth Online*, 17 July, http://planetearth.nerc.ac.uk/news/story.aspx?id=1024.

Kruger, H., Thompson, L., Clarke, R., Stenekes, N. and Carr, A. (2009) *Engaging in Biosecurity: Gap Analysis*, Australian Government Bureau of Rural Sciences, Canberra.

Latour, B. (2001) 'From "matters of facts" to "states of affairs": Which protocol for the new collective experiments?', www.bruno-latour.fr/sites/default/files/P-95-METHODS-EXPERIMENTS.pdf.

MAF (2011) *Statement of Intent 2011/2014*, Ministry of Agriculture and Forestry, Wellington, New Zealand.

MAFBNZ (2009) *Government Industry Agreements*, Ministry of Agriculture and Forestry, Wellington, New Zealand.

—— (2011) *Public and Industry Benefit in a Government/Industry Agreement*, Ministry of Agriculture and Forestry, Wellington, New Zealand.

Manzella, D. and Vapnek, J. (2007) *Development of an Analytical Tool to Assess National Biosecurity Legislation*, United Nations Food and Agriculture Organization, Rome.

Mumford, J. D. (2002) 'Economic issues related to quarantine in international trade', *European Review of Agricultural Economics*, vol. 29, pp. 329–48.

Owen, M. (2011) 'A strategy for tackling aquatic and riparian INNS', Third Sector GB Invasive Non Native Species and Biosecurity Conference, 7 June, www.theriverstrust.org/ seminars/archive/inns_june_2011/ART%20INNS%20-%2010%20Mark%20Owen.pdf.

Rothstein, H., Huber, M. and Gaskell, G. (2006) 'A theory of risk colonisation: the spiralling regulatory logics of societal and institutional risk', *Economy and Society*, vol. 35, no. 1, pp. 91–112.

Waage, J. K. and Mumford, J. D. (2008) 'Agricultural biosecurity', *Philosophical Transactions of the Royal Society B*, vol. 363, pp. 863–76.

Wilkinson, K. (2011) 'Organised chaos: an interpretive approach to evidence-based policy making in Defra', *Political Studies*, vol. 59, no. 4, pp. 959–77.

Wilkinson, K., Medley, G. and Mills, P. (2010) 'Policy-making for animal and plant diseases: a changing landscape?', Rural Economy and Land Use Programme, Policy and Practice Note 16, University of Newcastle, www.relu.ac.uk.

5

LEGAL FRAMEWORKS FOR BIOSECURITY

Opi Outhwaite

Introduction

Legal and regulatory measures are an important component of strategies for managing biosecurity risks. Adequate legislative provisions need to be in place at the domestic level to enable both preventative and responsive action to be taken, and these provisions must be successfully implemented and effectively enforced if they are to work. Whilst legal and regulatory interventions are not justified in every situation, measures which aim to prevent, contain or eradicate pests, diseases, invasive species and pathogens often depend upon a legal basis for action and, without this, scientific or political recognition of biosecurity issues may not be enough for biosecurity objectives to be met.

Adopting the necessary legal frameworks is not, however, always a straightforward exercise. Biosecurity is a complex and technical area and the development of effective legal frameworks presents a number of challenges. This chapter examines key aspects of the international legal framework for biosecurity and the challenges faced at the domestic level for countries seeking to implement or update relevant legal provisions and arrangements. In so doing, the chapter briefly highlights the historical development of domestic frameworks and the the development, nature and scope of key international instruments and bodies in the context of biosecurity.

Why is legislation important for biosecurity?

In many instances, effective responses to biosecurity risks rely on the existence of adequate legal provisions. As will be apparent from other chapters in this collection, there is a range of actions which a government may wish to take to prevent the entry into the country of pests, diseases, invasive species or pathogens, or to limit their spread or attempt to eradicate them. Such measures may apply to species themselves, to derivatives of those species (for instance, animal and food products) or to the vectors and pathways which facilitate their movement (for instance vehicles, packaging materials or farm equipment). In many instances these measures will apply to unintentional introductions or movement, but legislation may also establish a legal basis for intentional introductions (for instance, the intro-duction of exotic pests as part of a strategy for integrated pest management).

From a regulatory perspective, the classic approach to managing these risks revolves around prevention, eradication and control, with each stage tending to present progressively greater costs and risks (because of reduced likelihood of success and increased negative impacts). At the domestic level a variety of responses may be adopted in this context, and these can be usefully categorized as pre-entry, point-of-entry and post-entry controls.

For imported produce – a common source of introductions – measures might include, for instance:

Pre-entry

- prohibiting the entry of a commodity (for example a type of fruit or vegetable) which presents a high risk for the introduction of a particular pest;
- requiring that other imported commodities have been produced in an area that is free from a particular pest or have been subject to appropriate treatment to prevent the presence of the pest. This might in turn lead to measures for verifying compliance with these requirements, for instance through in-country (pre-export) inspections by authorities of the exporting country and/or the importing country.

Point-of-entry

- documentary and/or physical inspections of the commodities (or a sample thereof) at the point-of-entry, usually a border point;
- requiring treatment, return to origin or destruction of the commodity in the event of non-compliance with requirements or detection of the pest.

Post-entry

- quarantine;
- surveillance and inspection to detect outbreaks of the relevant pest by a designated regulatory authority;
- surveillance and/or reporting obligations for relevant stakeholders (for example, those most likely to come into contact with the pest);
- treatment or destruction in the event that the pest is detected;
- the imposition of sanctions and penalties in the event of non-compliance with obligations or restrictions.

In each case the powers of agencies and inspectors, the range of actions permitted and the circumstances under which action can be taken must have a legal basis. If not, then the action could result in a legal challenge or, for instance, in trade sanctions (see further the role of the World Trade Organization, below).

Legislation must therefore provide sufficient powers to enable action to be taken. It must be flexible and broad enough in scope to enable new risks to be addressed, but specific enough (including through the use of secondary legislation) to ensure that all parties (regulators and regulatees) are clear about the measures and the circumstances in which they apply. Provision for broader supporting arrangements, such as designation of approved laboratory facilities or authorized treatment, must be established in addition to the nomination of regulatory powers and responsibilities. The challenges posed by these requirements are discussed further below.

The development of legal controls for agriculture and public health

Although modern biosecurity concerns require the adoption of extensive legal measures, attempts to address aspects of biosecurity through legislative means are not new. In particular, the components of 'agricultural biosecurity' – plant health, animal health and food safety – have been subject to national and international control for some time.

In the case of plant health, attempts to address specific threats have taken place for over a century. In the UK, the Destructive Insects Act was adopted in 1877 and aimed to prevent the introduction and establishment of the Colorado Beetle (*Leptinotarsa decemlineata*) (Mumford et al., 2000). Historical measures such as these paved the way for the adoption of legal restrictions to minimize the risk of introducing pests and diseases from other geographical areas and for acting to contain or eradicate such pests in the event of an outbreak.

Ebbels (2003) suggests that the first legislative measures aimed at controlling a plant pest were in fact those taken by France and the USA, and later Germany, concerning the destruction of the common barberry (*Berberis vulgaris*) in the seventeenth and eighteenth centuries. As Ebbels explains, once the scientific basis for action had been established, legislation requiring the destruction of the barberry, or enabling local measures for its destruction, was introduced in several European states. The measures sought to address a specific plant health risk, though it was not until around 200 years later that the nature of the pest was understood. The efficacy of the law depended on the way in which it was framed. In some cases the extensive use of provisos and exceptions meant that it was virtually useless. In other cases where the requirements were simple, with few exceptions, and where supported by adequate enforcement, the legislation was very effective. In some countries such as the UK, where barberry had been less important, eradication was left as a voluntary action; in cases such as these there are apparently no records indicating a noticeable beneficial effect resulting from voluntary action (Ebbels, 2003).

Aspects of animal health (at this point referring principally to livestock) have also been regulated for centuries, with controls in this field again preceding an understanding of the disease agents and their epidemiology (Waage and Mumford, 2008). In the UK for instance, a veterinary health service has been in place since 1865 when a temporary department was established to respond to a rinderpest outbreak (DEFRA, 2011). The Cattle Diseases Act 1866 expanded upon the Sheep Pox Acts of 1848 and enabled authorities to impose movement restrictions and to order the destruction of infected animals (see also Spinage, 2003). Historical measures such as these were often more limited than those seen in modern animal health frameworks. Woods (2004a, 2004b) notes, for instance, that in the eighteenth and nineteenth centuries animal health controls in England tended to be limited to the enforcement of quarantine in the event of major epidemics. Such measures were unpopular and often not effectively enforced. Consequently legal interventions played a limited role and tended to address specific issues related to livestock and public health.[1] However, animal health and state veterinary services continued to develop substantially through the twentieth century.[2]

These examples illustrate the piecemeal, reactive and sectoral nature of legal responses to aspects of biosecurity. The focus here is also on agricultural 'biosecurity' and in particular on responses to economic threats associated with the agriculture sector as well as on major public health issues.

The international legal framework for biosecurity

While this brief description is only illustrative and focuses on the United Kingdom, the international framework largely reflects this historical approach. The starting point in examining the international legal framework for biosecurity is therefore to understand that there is no single instrument for addressing biosecurity and no comprehensive approach has been adopted at this level. Instead a number of instruments and agreements are applicable.

International standard-setting bodies

The establishment of international bodies in these areas principally took place in the early twentieth century. The Office International des Epizooties (the OIE, sometimes referred to as the World Animal Health Organization) was established in 1924 with the aim of introducing international measures in order to protect animal health. The OIE produces the Terrestrial Animal Health Code (and an equivalent for aquaculture), which sets out international standards for animal health (OIE, 2011). Although the OIE has a general mandate to address animal health, it has tended to focus on diseases of livestock and those which would have direct economic implications, and not on wild animals. This is evident from the categorization of most notifiable diseases (e.g. as diseases of cattle, sheep, horses, etc.).

This approach of listing diseases also highlights the historical development of the framework, in contrast with modern perspectives which may recognize more explicitly that pathogens can be relevant to a range of species (including those of both kept and wild animals) and that they may cause a number of additional and indirect effects (Perrings et al., 2010).

For plant health, a group of European countries took action in 1881 to control the spread of grape phylloxera following the extensive damage caused to European vineyards after its accidental introduction from North America (Ebbels, 2003). The International Plant Protection Convention (IPPC) eventually followed in 1951. The current international plant health framework is based on the IPPC 1997 and on the numerous specific standards published by the IPPC secretariat.[3]

The Codex Alimentarius Commission (Codex) was established in 1963 by the UN Food and Agriculture Organization (FAO) and the World Health Organization (WHO). This body seeks to establish food safety standards in order to protect human health and was developed out of concern to harmonize food laws and standards, especially in the light of international trade. It also sought to address consumer food safety concerns, which were of an increasingly technical nature (WHO–FAO, 2006). Numerous Codex standards have been developed and may deal with all characteristics of a particular commodity or with a specific characteristic, such as Maximum Residue Limits for pesticides or veterinary drugs.

The World Trade Organization

The regulation of international trade is of course a key issue for biosecurity since international trade is one of the principal drivers for the movement of species, diseases and pathogens. Measures which might be used to protect against the arrival or spread of such things (such as those described above) might, in practice, have the result of restricting trade (see also Jack, 2009).[4]

The general principles of the WTO (such as non-discrimination and 'national treatment') apply to many national biosecurity measures just as they do to other domestic measures which affect trade. More specifically the Agreement on Technical Barriers to Trade and, in particular, the Agreement on the Application of Sanitary and Phytosanitary Measures (SPS Agreement), have direct implications for the adoption of domestic legal measures for biosecurity. The frameworks of the IPPC, Codex and the OIE have assumed even greater significance in this context because they are formally recognized in Annex A(3) of the SPS Agreement and the standards and guidelines issued by them are assumed to be compatible with it and to reflect the general principles of the WTO. Although countries are free to adopt their own standards they must, in that case, be able to justify them based on risk analysis.

This framework clearly limits the types of measures that can be implemented by member countries, but this limitation is to be balanced against the perceived benefits of participation in the multilateral trading system. Member countries must, however, be mindful of the implications of these rules and standards on domestic biosecurity measures.

As already noted, some countries have been addressing particular biosecurity concerns for decades. Consequently, in some of these countries legislation was established long before the development of the modern multilateral trading system. Legislation in developing countries often dates to colonial times or was established upon independence (see also Ikin, 2002). One effect of the WTO framework is that member countries have needed to (or still need to) revise relevant domestic legislation. Whereas a country previously legislated according to its own needs and priorities, that legislation should now reflect WTO obligations. As well as being justified on the basis of risk, transparent, non-discriminatory and least-trade-restrictive technical terms and requirements should also be incorporated into legal measures. This may mean changing the content of the legislation – revising definitions, for instance, and introducing institutional changes (such as the designation of official contact points) – as well as changing the subject and nature of the restrictions that are permissible.

One of many examples of such revisions can be seen in the plant health law of Mauritius. Until relatively recently, the key statute was the Plants Act 1976. This Act did not reflect modern plant health requirements (in line with international standards and principles) in a number of ways. Key terms (as established by the IPPC) such as 'quarantine pest', 'phytosanitary certificate' and 'pest risks analysis' were not defined in the Act. 'Officers' were empowered to detain, examine, remove, treat, destroy, (etc.) any article which they had 'reasonable grounds' to suspect may be 'infected'. Such actions do not reflect principles of transparency and risk analysis. There were no provisions for the use or issuing of phytosanitary certificates and no schedules of pests subject to control. Whilst this legislation may or may not have served Mauritius well with respect to the country's own plant health concerns, it did not reflect international requirements in the light of the WTO 1994, though Mauritius became a member of the WTO as of 1 January 1995. The Mauritius Plant Protection Act 2006 sets out a revised framework including new definitions which reflect IPPC standards, such as a revised definition of 'pest' and new definitions for 'phytosanitary certificate', 'quarantine' and other terms which had been missing. A National Plant Protection Office was established and other measures enacted, such as powers to designate 'pest free areas'. A list of quarantine pests and powers to revise the list have also been adopted.[5] Although there is no obligation to revise 'pre-1994' legislation *per se*, a failure to do so may mean that it does not give effect to technical requirements set out in international standards or to broader WTO principles. In turn this may damage trade opportunities or even lead to a WTO dispute.

With respect to the nature and extent of measures that may be imposed, disputes have arisen between member countries regarding the 'appropriate level of protection' and the application of key concepts including scientific justification and precaution, in the context of the requirement for risk analysis. These disputes demonstrate that measures such as import bans and detailed sanitary or phytosanitary requirements can and will be challenged at the WTO.

One high-profile case was the 'Australia – Apples' dispute.[6] Australia banned the importation of New Zealand apples in 1921 following the entry and establishment of fire blight in New Zealand in 1919. New Zealand had sought to gain access to the Australian apple market from 1986, eventually initiating a dispute through the WTO dispute settlement process. New Zealand claimed that the Australian phytosanitary measures were not justified, challenging the import risk management measures that were then being imposed by Australia and the methodology on which the risk assessment was based. One of the arguments raised by Australia was that the measures were justified on the basis of the *available* evidence. The Dispute Settlement Panel found that the import risk analysis was not supported by scientific evidence and that less-trade-restrictive measures were available. On appeal, the Appellate Body also found that the measures were not justified. Australia subsequently replaced the existing measures with detailed controls and quarantine requirements.

Environmental protection and multilateral environmental agreements

Discussion of the environmental aspects of biosecurity, including matters related to biodiversity and ecosystems, and particularly to the management of Invasive Alien Species (IAS), is largely overlooked in the discussion above and this reflects the greater emphasis that had been afforded to 'agricultural biosecurity', including the development of the applicable legal instruments. As indicated, early measures often sought to address risks to livestock and crops because of the economic implications of losses in these areas.

This focus was initially reinforced by the FAO, which played a role in establishing the international standard-setting bodies and which continues to provide support to them.[7] The FAO has also played an important role in promoting the concept of biosecurity, emphasizing both the limitations that can arise, particularly for developing countries, in continuing to regulate on a sectoral basis and the benefits to be derived from a more integrated and strategic 'biosecurity' approach. It has, consequently, been an important actor in shaping the biosecurity agenda, but its earlier work on biosecurity had a narrower focus, framing biosecurity principally in terms of the three main sectors – food safety, plant health, and animal health and life. This focus has expanded, and in their 2007 'Biosecurity Toolkit', the FAO expanded their earlier definition, describing biosecurity as:

> a strategic and integrated approach to analysing and managing relevant risks to human, animal and plant life and health and associated risks to the environment ... Thus biosecurity is a holistic concept of direct relevance to the sustainability of agriculture, and wide-ranging aspects of public health and protection of the environment, including biological diversity.

In addition to this expanded FAO definition, the relevance of pests, diseases and pathogens in an environmental context is now more widely recognized and the nature of such risks – including the fact that they often do not respect regulatory or political boundaries – is better understood. The cross-cutting issue of IAS has also gained prominence, particularly since IAS are now recognized as one of the major drivers of global biodiversity loss (Jay et al., 2003; Meyerson and Reaser, 2002).[8] There is now, it can be argued, greater acceptance of biosecurity as a broader environmental concern.

Although none of the international environmental agreements addresses biosecurity expressly, a number of instruments in environmental law are relevant. FAO-Norway cites the following non-exhaustive list of 'sectoral instruments' related to biosecurity, which includes several multilateral environmental agreements (MEAs): the Rotterdam Convention on the Prior Informed Consent Procedure for Certain Hazardous Chemicals and Pesticides in International Trade; the Convention on Persistent Organic Pollutants; the FAO International Code of Conduct on the Use and Distribution of Pesticides; the Biological and Toxin Weapons Convention; the FAO International Code of Conduct on Responsible Fisheries; the Ramsar Convention on Wetlands; the Protocol to the Antarctic Treaty on Environmental Protection; the Convention on the Conservation of Migratory Species of Wild Animals; the Global Programme of Action for the Protection of the Marine Environment from Land-Based Activities. These agreements address specific environmental issues and in doing so impose obligations or guidance on implementing countries which may affect their biosecurity strategies. For instance, controls related to the presence of IAS may be required for the conservation of specific habitats such as wetlands.[9]

The Convention on Trade in Endangered Species (CITES)[10] is also important because it concerns international trade which, as we have seen, is a key pathway for the global spread of species, pests and pathogens. The significance of CITES is, however, limited insofar as it aims to protect *endangered* species by preventing their illegal trade and not to limit trade for the purpose of protecting species more generally. Nevertheless, articles of the Convention place further obligations on contracting states by introducing an import and export permit system for the movement and trade in specimens listed in the appendices. For example, Article III (2) requires that any species listed in Appendix I (species that are the most endangered) shall only be exported if accompanied by an export permit. Consequently, states that have ratified CITES must build these obligations and requirements into domestic measures, which place controls on the trade of plants and animals.

The most obviously and widely applicable of the environmental agreements is the Convention on Biological Diversity (CBD).[11] The CBD applies not only to general conservation measures but also to the specific components of biodiversity and to processes and activities, regardless of where the effects occur. In addition, Article 8 sets out obligations for *in situ* conservation, including:

- the establishment or maintenance of means to regulate and manage the risks associated with LMOs (living modified organisms) and biotechnology 'which are likely to have adverse environmental impacts that could affect the conservation and sustainable use of biological diversity, taking also into account the risks to human health' (paragraph g);

- preventing the introduction of, controlling or eradicating, alien species 'which threaten ecosystems, habitats or species' (paragraph h); and
- developing and/or maintaining legislation and regulatory provisions for the protection of threatened species and populations (paragraph k).

The Cartagena Protocol to the CBD sets out further measures relating to the regulation of LMOs. The protocol requires 'advance informed agreement' between exporting and importing states for the movement of LMOs intended for release into the environment and makes it clear that the precautionary approach should be adopted.

Obligations set out under the CBD will therefore have some impact on the way in which the CBD's implementing countries can regulate and manage biosecurity frameworks at the national level. In implementing the CBD, contracting parties must consider the wider context in which measures adopted may affect biodiversity. This requirement may have important consequences in terms of biosecurity regulation by shaping the way in which certain risks should be managed and the factors that must be taken into account when certain biosecurity decisions are taken. Most basically there is an obligation to ensure that measures adopted pursue the aims of preserving biodiversity, and this may require restrictions aimed at preventing the entry or spread of pests (etc.), which would have adverse affects.

Invasive species

The risks associated with the introduction into an area of 'alien' or 'non-native' species that spread and establish are now widely recognized. Detailed legal frameworks to prevent or manage such introductions are, however, less well established than those pertaining to certain other aspects of biosecurity. As seen above, the CBD includes a general obligation relating to the control of IAS, but there has been little detailed guidance as to how implementing countries should manage IAS threats within a domestic framework.[12] Influential guidelines have been developed by the International Union for the Conservation of Nature (IUCN, 2000), and the CBD later adopted its Guiding Principles for the Prevention, Introduction and Mitigation of Impacts of Alien Species that Threaten Ecosystems, Habitats or Species (UNEP/CBD/COP/6/23). These principles require, amongst other things, the adoption of a precautionary approach, the implementation of appropriate border controls and quarantine measures, and adoption of appropriate regulatory measures for both intentional and unintentional introductions in domestic frameworks.

Recognizing the limitations in this area, the CBD recently published the draft document, 'Considerations for implementing international standards and codes of conduct in national invasive alien species strategies and plans' (29 November 2011). Importantly, the document recognizes the role that is, or could be played, by other key instruments and organizations such as the OIE, IPPC, Codex and CITES. It recognizes also that the international legal framework is simultaneously overly complex, with several overlapping agreements, and inadequate in its scope, with numerous pathways for introduction not addressed by any of the applicable instruments (see also decision VIII/27). In addition it is recognized that

> Relatively few countries have invested in a comprehensive 'biosecurity' approach that addresses IAS through well-coordinated policies and pro-grammes across relevant sectors (esp., agriculture, environment, fisheries, trade, transport, development assistance, defence, and energy). It is common for IAS efforts to be poorly coordinated among the appropriate ministries and stakeholders.

At both the international and domestic levels a comprehensive approach to IAS can be lacking. It can be seen from the brief discussion above that, in the context of international law, IAS are framed in terms of conservation and environmental protection. This again reflects the somewhat fragmented, sectoral arrangements at the international level. In England and Wales the emphasis in legislation is on preventing the deliberate release of IAS or, in some cases, preventing their arrival. This is a narrower and less comprehensive approach than seen elsewhere in the contexts of plant and animal health. As argued by Perrings et al. (2010), strategies for inspection, detection, eradication and control are needed here. Further, whilst environmental legislation and enforcement agencies may not address wider issues related to IAS, the equivalent measures and bodies in other sectors such as plant and animal health may not have the authority to address IAS matters which are outside of the scope of their own work.

The increased recognition both of the problems of IAS and of the limitations in international and domestic legal capacity is significant, but the present position remains that there is no binding international agreement applicable generally to IAS.

A review of the international legal framework makes it clear that, whilst there are a number of standards and obligations which affect the types of measure that a country can adopt, there is also no overarching approach or agreed set of principles with respect to biosecurity. In addition there are a number of gaps and potential conflicts within this framework that provide an inconsistent basis for action.

Implementing domestic legal frameworks for biosecurity

It can be seen from the overview above that the historical development of frame-works for health and environmental protection, coupled with numerous limita-tions of the applicable international instruments, give rise to a number of difficulties in the implementation of effective legal controls for biosecurity.

First, domestic arrangements are often still organized around traditional sectoral divisions (plant health, animal health, food safety and conservation). These divisions reflect piecemeal developments, which took place before the nature of the risks involved was well understood. Consequently they may not represent the most effective means of managing the pathways by which diseases, pests or species enter into an area and are moved from one location to another. The arrangements may also give rise to legal and regulatory 'gaps' which prevent disease risks (for instance) from being effectively managed.

Second, there remain numerous obstacles within the international framework. In this case sectoral divisions are again in place. Tensions between various agreements also remain, particularly concerning the 'free trade' objectives of the WTO, which require measures to comply with international standards and be based upon risk, and the environmental protection and sustainability goals of most environmental agreements, which advocate or require a precautionary approach. The international standards produced by the IPPC, OIE and Codex do not provide a clear framework for the adoption of precautionary measures or broad-based risk responses, whilst the precautionary principle remains a keystone of the Convention on Biological Diversity (and broadly of the Cartagena Protocol) and of other MEAs (Stilwell and Tarasofsky, 2001; Huei-Chih, 2007; Weiss, 2003; Trouwborst, 2007).

The tensions between these guiding principles for decision making are built upon the differing objectives of the agreements. The WTO and the international standards developed in relation to the WTO agreements reflect the overarching objectives of increasing international trade in the context of global economic liberalization. The MEAs seek to further the preservation of natural resources and biodiversity. Clearly this is not an entirely consistent basis from which to develop legal and regulatory strategies.

Inconsistency in the use of key concepts and terms at the international level also presents challenges. The term 'biosecurity' does not itself appear in any of the key agreements, having developed after most of these were drafted. This lack of international guidance is reflected in differences in the meaning of 'biosecurity' within domestic legal frameworks. In the UK, for instance, biosecurity is used mainly to describe 'on-farm' measures, such as disinfecting vehicles and regulating local movement. In the American context it has traditionally been linked with bioterrorism, including deliberate introductions of pests and diseases to attack food security, the agricultural economy or public health. In Australia and New Zealand, a more comprehensive approach, in line with the modern definitions described above, has been pursued (Donaldson, 2008; Hinchcliffe and Bingham, 2008; Jay et. al., 2003, Fletcher and Stack, 2008).[13] These approaches reflect different ideas about the processes, aims and outcomes of biosecurity, and in the first two cases do not necessarily reflect the integrated nature of biosecurity which is key to its relevance.

Other terms are also used inconsistently. For instance, terms such 'alien species', 'invasive alien species', 'introduction', 'precautionary approach', or 'risk analysis' are not used consistently by the IPPC and the CBD, despite the clear overlap and potential for coordination in this area. As noted, efforts to improve international coordination are being undertaken but a lack of consistency remains. Attempts by the IPPC to harmonize terminology related to IAS with the CBD had limited success, with the IPPC reporting that differing terms were based on different concepts and therefore had different meanings and could not easily be harmonized (IPPC, 2010). Again, this means that the basis for action at the domestic level can be unclear or contradictory.

Building on all of these limitations, particular biosecurity issues are in some cases entirely absent from applicable legal frameworks. Existing legal and institutional arrangements may be inadequate in the case of newly identified and emerging concerns, since these have not typically been subject to legal restrictions. For instance, declining honey bee (*Apis mellifera*) populations have attracted a great deal of attention worldwide (e.g. Benjamin and McCallum, 2008). Regulatory measures which minimize the spread of endemic or exotic pests and diseases of honey bees may therefore be an important component in the successful management of these populations. In England and Wales challenges arise because honey bee health does not fall neatly into any of the established regulatory paradigms – those for animal health, protection of pets or wildlife conservation – and is not subject to detailed measures which apply in some other contexts (for instance health of economically important livestock species). Honey bees are not regulated as livestock post-entry, though they are treated as such for the purposes of international trade. They similarly do not fall within the measures applicable to pets, and relevant risks are not covered by the domestic framework for environmental protection and invasive species.

More broadly, wildlife health in general may be overlooked in legal frameworks which, as noted, have developed principally to address either economic and agricultural objectives (i.e. in the case of animal health) or conservation objectives (for instance concerning the trade in endangered species). The trade in wildlife and wildlife products (such as bushmeat) again does not necessarily fall easily within existing regulatory paradigms, and important pathways for the introduction and spread of diseases – including zoonoses – might consequently not be subject to key legal interventions such as border controls and surveillance and reporting obligations. Smith et al. (2012) note for instance that, in the USA, the United States Department of Agriculture (USDA) does not have a general remit to regulate species as potential threats to wildlife or public health. Specific matters are addressed by a range of agencies, but this again does not provide a comprehensive approach. Waage and Mumford (2008) have identified a number of further pathways which are typically not addressed in biosecurity frameworks, including food aid, military operations and infrastructure development.

Even where comprehensive legal frameworks are developed, their success depends upon successful implementation and enforcement. With respect to biosecurity this can again present a formidable task. As discussed, in order to respond effectively to identified risks and to comply with international obligations, enabling legislation must be in place which provides sufficient powers to act in the face of biosecurity concerns. Numerous activities may be required, from preventing or restricting the entry of goods, imposing notification or reporting requirements in relation to specified pests, diseases or IAS, or requiring treatment or destruction where such a pest has been detected, to specifying practices to be followed and imposing sanctions for failure to comply with the law. This primary or enabling legislation is usually supported by secondary legislation in the form of decrees, regulations or orders giving further detail such as lists of notifiable diseases or restricted commodities.

The development of such legislation is in itself a substantial task given that the legislation must be up to date and may need to be amended frequently (particularly in the case of secondary legislation) so that it reflects current international standards and identified risks. As noted above, one difficulty in this respect is that legislation can often be outdated. This is a particular difficulty for countries which experience substantial bottlenecks when attempting to introduce new legislation. Outdated legislation means that action might not be taken when needed because there is no legal basis for it, or, if the authorities act anyway, that the action might not be legal (and consequently might be challenged by those who are subject to the measures in question) or in some instances might not be compatible with international obligations, including those under the WTO agreements.

In addition to adopting effective legislation it is essential that this law has 'real life' application; the law 'on the books' can itself do little if it is not appropriately implemented and enforced. This again can impose a considerable burden on implementing countries, not only (but particularly) those challenged by limited resources. The legislation in this case needs to be supported by appropriate institutional arrangements and by the designation and availability of regulatory authorities which are sufficiently empowered and practically able to act. Significant technical expertise may be required to support the measures, including pest risk analysis and diagnostic capacity as well as qualified and trained inspectors. Facilities and equipment for undertaking analyses and testing will be needed as well as the financial resources to obtain and maintain the necessary equipment. Overall, there are likely to be substantial requirements in terms of human, financial and technical resources. For many countries maintaining such frameworks can be difficult and there is an ongoing need to address commitments to capacity building in this area (see also Perrings et al., 2010).

Conclusion

The adoption of appropriate legislation and regulatory arrangements is an essential component of any biosecurity framework. Countries may wish to take a wide variety of measures in response to risks posed by pests, diseases, pathogens, IAS, contaminants, and so on, and will often be prevented from doing so if the under-lying legal arrangements are inadequate. The types of measure that can be adopted are influenced by a number of international obligations, standards and guiding principles. But problems can arise here because of a lack of clarity and consistency amongst the relevant instruments. In addition to this, domestic frameworks may already be organized according to historical arrangements which do not necessarily provide for an efficient, proactive and comprehensive approach. Finally, attempts to modernize or even to maintain these measures can be difficult in the light of the considerable challenges surrounding their implementation and enforcement. In many cases, therefore, there is considerable work to be done to ensure that legal frameworks provide an efficient and effective means of managing biosecurity risks.

Notes

1 See http://animalhealth.defra.gov.uk/about/aboutanimalhealth/history.html.
2 Food laws were also developed from the early nineteenth century (WHO–FAO, 2006).
3 International Standards for Phytosanitary Measures (ISPMs).
4 See Clive Potter, Chapter 8 in this volume, for further discussion of the WTO.
5 These authorize measures necessary to protect human, animal or plant life or health, or relating to the conservation of exhaustible natural resources.
6 Australia – Measures Affecting the Importation of Apples from New Zealand, WT/DS367/AB/R, adopted 17 December 2010.
7 See also Karki (2002) concerning food safety laws in countries within the South Asian Association for Regional Cooperation.
8 The International Plant Protection Convention was approved by the FAO conference (sixth session) on 6 December 1951, by resolution no. 85/51. The Codex Alimentarius Commission was established with support from the FAO and WHO (Codex 2006).
9 These shifts are reflected to an extent in developments in plant and animal health standards. For instance, changes to ISPMs have sought to clarify the extent to which environmental considerations can be taken into account when assessing plant health risks (see ISPM 11, supplement 1, section 2.3.1); see also FAO–OIE–WHO Collaboration, 'Sharing responsibilities and coordinating global activities to address health risks at the animal-human-ecosystems interfaces: a tripartite concept note', April 2010.
10 For example, resolution VIII.18 of the Ramsar Convention (invasive species and wetlands) urges parties to take decisive action to address the problem of IAS in wetland ecosystems, including through the application of risk assessment and incorporation of policies into domestic legislation.
11 Convention on International Trade in Endangered Species of Wild Fauna and Flora (Washington DC, 3 March 1973).
12 Convention on the Conservation of Migratory Species of Wild Animals (Bonn, 23 June 1979).
13 Where such guidance does exist it has been restricted to particular issues or ecosystems rather than being generally applicable. For instance, the International Maritime Organization's International Convention for the Control and Management of Ships Ballast Water and Sediments (the IMO Convention) aims to control the transfer of IAS through ballast water.

References

Benjamin, A. and McCallum, B. (2008) *A world without bees. The mysterious decline of the honey bee – and what it means for us.* Guardian Books, London, UK.

CBD (2011) *Considerations for Implementing International Standards and Codes of Conduct in National Invasive Alien Species Strategies and Plans* (29 November 2011), Secretariat of the Convention on Biological Diversity, United Nations Environment Programme, Montreal.

—— (2002) *Guiding Principles for the Prevention, Introduction and Mitigation of Impacts of Alien Species that Threaten Ecosystems, Habitats or Species* (UNEP/CBD/COP/6/23), Secretariat of the Convention on Biological Diversity, United Nations Environment Programme, Montreal.

DEFRA (2011) *Animal Health, A History*, http://animalhealth.defra.gov.uk/about/abouta nimalhealth/history.html.

Donaldson, Andrew (2008) 'Biosecurity After the Event: Risk Politics and Animal Disease', *Environment and Planning*, vol. 40, pp. 1552–1567.

Ebbels, D. L. (2003) *Principles of Plant Health and Quarantine*, CABI Publishing, Wallingford, UK.

Fletcher, J. and Stack, J. P. (2008) 'Crop Biosecurity: Definitions and Role in Food Safety and Food Security', in M. L. Gullino, J. Fletcher, A. Gamliel and J. P. Stack (eds), *Crop Biosecurity: Assuring Our Global Food Supply*, NATO–Springer, Dordrecht.

Grant, I. E. and Kerr, W. A. (2003) 'Genetically Modified Organisms and Trade Rules: Identifying Important Challenges for the WTO', *The World Economy*, vol. 26, no. 1, pp. 29–42.

Hinchcliffe, S. and Bingham, N. (2008) 'Securing Life: The Emerging Practices of Biosecurity', *Environment and Planning*, vol. 40, pp. 1534–1551.

Huei-Chih, N. (2007) 'Can Article 5.7 of the WTO SPS Agreement Be a Model for the Precautionary Principle?', *SCRIPTed*, vol. 4, no. 4, pp. 367–382.

Ikin, R. (2002) 'International Conventions, National Policy and Legislative Responsibility for Alien Invasive Species in the Pacific Islands', *Micronesia Supp.*, vol. 6, pp. 123–128.

International Union for the Conservation of Nature – World Conservation Union (IUCN) (2000) *Guidelines for the Prevention of Biodiversity Loss Caused by Alien Invasive Species*, IUCN, Gland, Switzerland.

IPPC (2010) ISPM 05 (Glossary of Phytosanitary Terms), appendix no. 1, IPPC Secretariat, Food and Agriculture Organization of the United Nations, Rome.

Jack, B. (2009) *Agriculture and EU Environmental Law*, Ashgate, Farnham, UK.

Jay, M., Morad, M. and Bell, A. (2003) 'BiosecurityK, A Policy Dilemma for New Zealand', *Land Use Policy*, vol. 20, pp. 121–129.

Karki, Tika Bahadur (2002) 'Sanitary and Phytosanitary (SPS) Measures in SAARC Countries', Discussion Paper. SAWTEE, Kathmandu and CUTS, Jaipur.

Meyerson, L. A. and Reaser, J. K. (2002) 'Biosecurity: Moving Toward a Comprehensive Approach', *BioScience*, vol. 52, no. 7, pp. 593–600.

Mumford, J. D., Temple, M. L., Quinlan, M. M., Gladders, P., Blood-Smyth, J. A. et al. (2000) *Economic Policy Evaluation of MAFF's Plant Health Programme*, Report to Ministry of Agriculture, Fisheries and Food, London.

OIE (World Organization for Animal Health) (2011) *Terrestrial Animal Health Code 2011*, available at www.oie.int/international-standard-setting/terrestrial-code/access-online.

Perrings, C., Burgiel, S., Lonsdale, M., Mooney, H. and Williamson, M. (2010) 'Globalisations and Bioinvasions: The International Policy Problem', in C. Perrings, H. Mooney and M. Williamson, *Bioinvasions and Globalisation*, Oxford University Press, Oxford.

Smith, K. M., Anthony, S. J., Switzer, W. M., Epstein, J. H., Seimon, T., Jia, H., Sanchez, M. D., Huynh, T. T., Gale Galland, G., Shapiro, S. E., Sleeman, J. M., McAloose, D., Stuchin, M., Amato, G., Kolokotronis, S.-O., Lipkin, W. I., Karesh, W. B., Daszak, P. and Marano, N. (2012) 'Zoonotic Viruses Associated with Illegally Imported Wildlife Products', *PLoS ONE*, vol. 7, no. 1, e29505, DOI: 10.1371/journal.pone.0029505.

Spinage, C. A. (2003) *Cattle Plague: A History*, Kluwer/Plenum, New York.

Stilwell, M. and Tarasofsky, R. (2001) 'Towards Coherent Environmental and Economic Governance: Legal and Practical Approaches to MEA-WTO Linkages', WWF–CIEL discussion paper, WWF International/CIEL, Gland, Switzerland.

Trouwborst, A. (2007) 'The Precautionary Principle in General International Law: Combating the Babylonian Confusion', *RECIEL*, vol. 16, no. 2, pp. 185–195.

UNFAO, (2007) *FAO Biosecurity Toolkit*, FAO, Rome.

Waage, J. K. and Mumford, J. D. (2008) 'Agricultural Biosecurity', *Philosophical Transactions of the Royal Society*, vol. 363, pp. 863–876.

Weiss, C. (2003) 'Scientific Uncertainty and Science Based Precaution', *International Environmental Agreements*, vol. 3, no. 2, pp. 137–166.

WHO–FAO (2006) *Understanding the Codex Alimentarius*, available at www.who.int/foodsafety/publications/codex/understanding_codex/en/index.html.

Winter, G. (2003) 'The GATT and Environmental Protection: Problems of Construction', *Journal of Environmental Law*, vol. 15, no. 2, pp. 133–140.

Woods, A. (2004a) *A Manufactured Plague: The History of Foot-and-mouth Disease in Britain*, Earthscan, London.

—— (2004b) 'The Construction of an Animal Plague: Foot and Mouth Disease in Nineteenth-century Britain', *Social History of Medicine*, vol. 17, no. 1, pp. 3–39.

WTO (1998) *Understanding the WTO Agreement on Sanitary and Phytosanitary Measures*, available at www.wto.org/english/tratop_e/sps_e/spsund_e.htm.

6

BIOSECURITY

Whose knowledge counts?

Gareth Enticott and Katy Wilkinson

Introduction: reframing animal disease

Studies of the interactions between scientists, society and the environment show that different people understand problems in different ways using different forms of expertise and knowledge. For policy makers, recognizing these differences is important for they affect the way science is understood and acted upon. In studies of science, technology and the environment, the classic example is provided in Brian Wynne's (1992, 1996) account of the British government's reaction to radioactive fallout in Cumbria following the Chernobyl disaster. Scientists descended on farms to offer technical advice to farmers but failed to grasp how farmers' practical knowledges were interwoven with their own identities. Ignoring these complex relationships can often mean that traditional forms of science communication fail to address the problems at hand, and sometimes make them worse.

Being open to different forms of knowledge and expertise has much to offer attempts to understand and improve biosecurity practices. However, attempts to manage animal disease are rarely framed in this nuanced fashion. Rather, the approach to managing and encouraging biosecurity has preferred top-down forms of regulation and/or communication of scientific evidence in the expectation that farmers will change and adopt new biosecurity practices. In this chapter, we pay closer attention to the ways in which animal disease is understood by different actors, how different knowledges are produced and what happens when contrasting styles of knowledge production and practice meet. Our argument is that, without finding ways to accommodate different perspectives on animal disease, attempts to resolve biosecurity issues will struggle. We begin by providing a brief historical overview of approaches to managing animal diseases before focusing on two recent high-profile diseases in the UK: the outbreak of foot-and-mouth disease (FMD) in 2001, which resulted in the slaughter of 10 million animals and cost £8 billion to manage (Anderson, 2002); and bovine tuberculosis (bTB) – a disease endemic to cattle in parts of England and Wales and whose management is complicated by the involvement of badgers – a protected species but also a disease vector (Enticott, 2001). Using these two examples, we chart what sorts of knowledges have proved influential in shaping animal disease policies before outlining how alternative forms of knowledge and expertise have challenged them. As a way through these apparently conflicting positions, we then introduce the concept of interdisciplinarity to suggest a more productive way of organizing knowledge of animal disease.

Biosecurity and the emergence of veterinary expertise

In 2011, two important biosecurity landmarks were celebrated: the 250th anniversary of the formal establishment of the veterinary profession and the eradication of cattle plague (known as rinderpest). It was a neat coincidence, for it was the discovery of rinderpest that led to the creation of a new science – veterinary science – to manage it. The first steps to manage the disease were taken by Pope Clement XI in 1711 who, on the recommendation of his physician Giovanni Lancisi, instructed the slaughter of all infected and exposed cattle. Farmers who objected were hung. Unsurprisingly, the policy was not popular, but rinderpest was eradicated. This policy of 'stamping out' animal disease was brought to Britain in 1714 when rinderpest struck the cattle population. This time, it was King George I's surgeon Thomas Bates who instigated a culling policy, with financial compensation provided to farmers to offset their opposition.

Not only did Bates and Lancisi succeed in ridding areas (at least temporarily) of rinderpest, but this policy of 'stamping out' as it became known was institutionalized as a fundamental logic of biosecurity within veterinary science. Yet, this way of organizing animal health only became possible thanks to a powerful and mutually constitutive relationship between the veterinary profession and the state during the nineteenth century. Until then, farmers accepted many diseases, such as foot-and-mouth disease, as unfortunate, but unavoidable, occurrences. Those who pressed for animal disease regulations received little support, certainly from those in the meat trade who feared for their livelihoods. The veterinary profession itself had limited experience or understanding of common diseases of livestock or their implications for public health. Instead their expertise lay largely in equine medicine; cattle work paid less and was not a significant part of veterinary training (Woods, 2004; Waddington, 2006).

However, during the mid-1800s powerful landowners calculated that the costs incurred from disease outbreaks were high and lobbied Parliament for legislation to prevent its spread (Perren, 1978). Veterinary leaders began to realign the profession with a wider public health agenda by asserting vets' credentials to speak on matters of animal health. They were assisted by the state, keen to exert control over the productive capabilities of its population. What emerged was a vision of animal disease as a national problem requiring state intervention for the sake of public health, farming productivity and animal health more generally. Associating veterinary science with agricultural productivism led to the establishment of veterinary laboratories to conduct experiments and develop diagnostic tools and vaccines. The government funded and organized veterinary education, advisory services for farmers and field veterinary services, and established a national State Veterinary Service. Vets directly benefitted from greater employment opportunities within government and regular government-funded work in private practice. The veterinary profession was therefore able to claim authority over animal health, and that authority lay in the increasingly scientific expertise of veterinary medicine which flourished under state sponsorship (Worboys, 1991).

The relationship between the state and the discipline of veterinary science had many effects on the way animal disease was understood. First, it led to an objective classification of the types of animal diseases that required intervention. This classification of 'notifiable' diseases – such as brucellosis, bovine tuberculosis, and foot-and-mouth disease – reflected not only the relationship between agricultural interests, vets and the state but also the objects of veterinary expertise. Some animal diseases were to command veterinary attention because of their economic or public health impacts (Hardy, 2003; Waddington, 2004; Woods, 2009), whilst others were left to farmers to deal with themselves (Woods, 2007). These disease classifications continue to direct the activities of veterinary science even when questions over the continued need to control some diseases are raised (Waage and Mumford, 2008).

Second, resources were allocated to investigate scientific solutions to the problems of animal disease. Whilst the benefits of stamping out policies for exotic diseases like FMD were quickly realized, the same could not be said for endemic diseases like bovine tuberculosis and brucellosis. For these diseases, improved diagnostic procedures were required that could only be developed within scientific laboratories. Woods (2011, p. 1944) argues that until the mid-nineteenth century vets had 'taken little interest in research owing to their faith in stamping out. Now, however, they viewed it as a way of making stamping out possible'. This reinforced the veterinary profession's claims to 'scientific' status. Rather than simply stamp out disease, the scientific status of the veterinary profession made it valuable to government policy makers (Hardy 2003; Waddington 2004). Thus, when the Ministry of Agriculture and Fisheries was established in 1919, vets were able to dominate policy making (Woods, 2011).

This was witnessed most vividly in the examples of bTB and FMD. For bTB, rising incidence of the disease during the 1980s and 1990s in the UK led to a scientific review of policy in 1997. The aim of the review was to address whether badgers were responsible for its spread. Evidence on whether culling badgers could eliminate the disease was inconclusive. As a result, the review suggested a culling trial using a standard scientific methodology – a randomized controlled trial – where badgers would be culled in some areas, and left alone in others (Krebs et al., 1997). The trials were to be run by an Independent Scientific Group (ISG) funded by but separate from government in order to ensure validity and objectivity. In 2007, the ISG issued its final report, suggesting that badger culling was unlikely to offer a meaningful solution to the problem of cattle TB (Independent Scientific

Group, 2007) primarily because of the phenomena known as 'perturbation' – a process by which badger culling disturbs established badger territories and encourages migration and disease spread, thereby offsetting any benefits gained by eradicating badgers from the original infected area (Woodroffe et al., 2005). Instead, the ISG recommended that greater testing of cattle for bTB and the use of more specific diagnostics should be deployed.

The outbreak of FMD in 2001 was a different challenge: whilst bTB is a relatively slow moving disease, fast moving exotic diseases like FMD must be stamped out quickly lest they become endemic. Nevertheless, the solution to this animal disease crisis still lay in scientific expertise. As the disease spread rapidly across the country, epidemiological modelling provided the government with information that could be used to guide its policy actions. One of the teams modelling the disease proved to be particularly influential. Relying on a limited dataset, they were able to show that, by culling all animals within a radius of 3 km from each infected farm, the disease could be brought under control. This policy, known as contiguous culling, would go on to be implemented, but at a high cost: 10 million animals would be slaughtered, most of which were healthy (Campbell and Lee, 2003).

Contesting biosecurity expertise

The experiences of FMD and bTB show the primacy of scientific expertise in biosecurity controversies. But scratch deeper, and the fragmented and contested nature of animal health knowledge and the nature of veterinary expertise itself is revealed. This should be of no surprise: studies of science and technology remind us of the constructed nature of scientific expertise. Whilst uniformity, certainty and irreversibility are characteristics of scientific knowledge, ensuring scientific facts can travel through time and space (Latour, 1988), the extent to which the practice of science itself reflects such unvarying standards is open to question. In practice, technology and science are often unruly: comprised of constantly evolving practices to fit contingent situations in order to make sense of phenomena that deviate from expected or standardized practices (Wynne, 1988). This requires a form of expertise and understanding that is in stark contrast to that associated with simple rule-following, such as scientific protocols (Berg, 1997). It is this ability to cope with uncertainty that characterizes advanced forms of expertise (Dreyfus and Dreyfus, 1986). Here, expertise is formed within communities of practice (Brown and Duguid, 1991) in which knowledge and practice is highly localized, reflecting local practices that are jealously guarded through rhetorical practices of professional 'boundary work' (Jasanoff, 1987). These activities reveal two further aspects of knowledge. First, that scientific expertise consists of a range of 'styles of reasoning' (Hacking, 1983) which describe different forms of knowledge practice and the divides and distinctions between them. Recognition of expertise is therefore derived from peers and clients, based on working relationships, mutual understanding and trust. And second, that along this continuum of what constitutes expertise, forms of practical knowledge, judgment and skill will reside within other professions that rely on embodied and intuitive forms of expertise, as opposed to the scientific world.

How do these concerns relate to biosecurity? Whilst it is clear that veterinary science grew in stature throughout the nineteenth and twentieth centuries, and the credibility of the veterinary profession with it, Woods (2007) points out that tacit knowledge and embodied practices were crucial to the development of animal health techniques such as pregnancy diagnosis, and with them the shape of the veterinary profession too. More recently, however, these forms of veterinary field knowledge appear to be have been marginalized in the control of animal disease. First, the FMD crisis highlighted a clash in styles of expertise (Bickerstaff and Simmons, 2004). A key moment in the management of FMD was the decision to follow the advice of epidemiological modellers rather than the advice from field-based vets in the State Veterinary Service. Arguably this moment revealed the distinctions between two different forms of expertise (Bickerstaff and Simmons, 2004). On the one hand, the modellers offered a 'distant' form of knowledge, whose certainties were based on a formalized but generalized view of the farmed environment. On the other, state vets' 'proximate' form of knowledge offered an alternative view that was more finely tuned to local geography and farming practices. That this distant form of knowledge was in keeping with the Labour government's command-and-control style of governing, combined with their loss of faith in the ability of field vets to resolve the problem (Ward et al., 2004), led to the mass slaughter of millions of farm animals.

Although the experiences of FMD suggest that formalized and generalized expertise displaces that which is locally situated, other research suggests that the latter is never completely marginalized. Indeed, just as studies show that the implementation of policy or regulations is shaped by social judgments (Lowe et al., 1997), so is it possible to see the same social processes creating knowledge and expertise in the practice of biosecurity. This is clearly demonstrated in the practices that vets use to detect bTB. In the UK, private vets are mostly responsible for conducting bTB tests, for which they must follow a strict protocol laid down in European law. However, recently farmers have complained that some vets were not testing according to these rules, and government investigations confirmed these suspicions. As a result, the government has sought to remind and educate vets of the importance of following the standardized testing protocol and has suspended those who do not.

A more effective strategy, however, would be to understand why variations in protocols occur in the first place, for it may be that permitting degrees of flexibility is central to the workings of standardized knowledge (Timmermans and Berg, 1997). Ethnographic studies of vets reveal how they are immersed in a range of social and natural relations that together construct forms of expertise that allow a standardized test to be implemented in practice (Enticott, 2012). In short, these alternative forms of expertise allow vets to 'get the job done'. In this case, the constant routine of bTB testing means many vets develop embodied and tacit skills. For example, injecting cows throughout the day can lead to the recognition of tactile sensations in the vet's hand that suggest the injection was successful without having to inspect by hand as the protocol demands. These skills are particularly important when working in difficult and dangerous conditions. The routine of testing also alerts vets to the limitations and exceptions of the bTB test, and that it should be understood in relation to the local conditions rather than as a universal standard. Cattle whose infectious status is borderline may be re-examined again and again to ensure a 'correct' diagnosis; cattle from one part of a farm may be treated differently; and the picture of the disease on the farm will inform the way test results are interpreted.

Veterinary biosecurity practices are also shaped by the social nature of veterinary work. Research on how people cope with complex and uncertain conditions as part of their work reveals that social learning plays a significant role in shaping people's practices (Orr, 1996). By sharing experiences about coping with uncertainty as part of work, these stories shape work practices as well as conferring a group identity or belonging to a community of practice. In the same way, working amongst other vets helps prepare vets for the practice of veterinary medicine and establishes shared goals and identities. Achieving this identity is central to learning how to become a vet and is often expressed by the sharing of tips and stories of ways around problems when, for example, it is difficult to follow rules or protocols. Vets, working on their own and in challenging conditions will often seek advice from their colleagues at the end of the day. Telling stories of how these were overcome during bTB testing helps establish an heroic quality to the vet; leads to the classification of bTB tests according to whether they are good (i.e. quick) or bad (i.e. slow); and legitimizes alternative forms of expertise with which to conduct bTB tests. The idea of the 'good' vet and a 'good' test is also legitimized by farmers. Young graduates may especially feel these pressures as they seek acceptance into a professional culture. As they want to move on to other aspects of large animal practice, they must first gain the acceptance of local farmers. Departing from these accepted practices may therefore affect a vet's future career opportunities within the farming community.

Disputes over biosecurity knowledge and practice are not just limited to debates between scientists and veterinary surgeons. Farmers themselves have their own sources of practical experience and cultural beliefs that shape their understanding of disease risks and help them weigh up the benefits of biosecurity interventions. In relation to FMD, Heffernan et al. (2008) describe how the acceptance of a vaccination programme in Bolivia relied on a process of reinvention in which farmers reconceived the way vaccination worked to fit with their own cultural notions of disease. Uptake was therefore not due to scientific or economic arguments, but vaccination discourses were 'reinvented' to fit in with local beliefs. This included the belief that FMD was caused by heat so vaccination worked by 'cooling' animals. Thus, Heffernan et al. (2008, p. 2439) argue that 'farmers were not vaccinating against the disease threat itself, but rather the imbalances of hot and cold, underlying the disease process'.

In the UK, research suggests that farmers make decisions about biosecurity in a similar way to which people make decisions about their own health. In public health, healthy behaviour is shaped by competing sets of practical priorities; the day-to-day experience of health risks; and cultures of risk that define the kind of life (and risks) worth taking (Backett, 1992; Lupton and Tulloch, 2002). These forms of 'lay knowledge' about health challenge the identification of risk factors by epidemiologists and the preventive actions they recommend (Davison et al., 1991).

Similarly, research into farmers' understandings of bTB and biosecurity (Enticott, 2008) suggests that farmers construct their own understandings of disease and the validity of preventive measures through their own and their peers' experiences of living with disease. To make sense of bTB, farmers construct 'candidates': ideal types of cows, farms and farmers who are likely to suffer from bTB. These candidates are based on their first-hand experiences of the disease, which are shared within farming communities suffering from the disease. Using this candidate system helps farmers to make sense of disease: it helps them retro-spectively explain why they have suffered a bTB breakdown and/or to predict who is likely to get bTB based on such things as other farmers' management practices or their farming ability.

Although these rules are often similar to generic scientific advice on biosecurity distributed by the government, the candidate system that farmers use accommo-dates exceptions to these rules, such as 'good farmers' suffering from bTB break-downs. Farmers point to these exceptions as 'unwarranted survivals' or 'unwarranted deaths' of cattle from bTB, such as in cases where cattle-to-cattle transmission was likely to occur but did not, or examples of low-risk 'closed herds' contracting bTB. Faced with the generic biosecurity advice offered by the govern-ment, these experiences instead come to characterize universal systems of animal health knowledge as fallible and dependent on luck. Set against these exceptions, universal biosecurity knowledge appears to inspire a sense of fatalism amongst farmers whereby they believe nothing could be done to prevent animal disease. This more nuanced understanding of disease and its uncertainties – in part based also on the practice of farming and a distrust in government advice resulting from previous animal health scares, such as FMD – means that farmers create and rely on their own 'lay epidemiologies' of disease management. In practice this means missing or delaying bTB tests; ignoring biosecurity regulations, such as isolating bTB infected cattle; illegally killing wildlife suspected of spreading disease; and basing cattle purchasing decisions on their own ideas of stress and immunity.

Broadening the evidence base: the role of interdisciplinarity

What role should these different forms of knowledge and expertise have in the management of animal disease? As the UK government pursues an agenda of 'responsibility and cost sharing', whereby farmers must take ownership of more areas of disease control and government pays less compensation for affected livestock, it is becoming increasingly important to understand how farmers perceive and respond to disease risk. Moreover, rather than insist on the primacy of scientific knowledge in animal health disputes, we might be better off developing new types of biosecurity practices that are more open to uncertainty and knowledges which are somehow 'more than' scientific.

An obvious method for developing new forms of biosecurity practices and more-than-scientific knowledges is by strengthening the number of social scientists and economists in government, and seeking external advice from non-scientists. In the UK, social science advice has, until very recently, been almost non-existent in government departments concerned with biosecurity, such as Defra. As a department concerned with agriculture and the environment, it saw itself primarily as a consumer of scientific and veterinary expertise. The non-scientists that did exist consisted mainly of economists. The impact of FMD has, however, led to change: the creation of the Science Advisory Council (SAC) in 2004 included a social science subgroup that reported on the state of social science in Defra and called for more social scientists and a broader range of evidence to be included within the policy-making process (Science Advisory Council, 2006). However, the recent creation of a social science experts panel, working jointly for Defra and the Department for Energy and Climate Change (DECC), confirms the continued separation, and thus potential marginalization, of social scientists within the policy process.

A more significant and lasting solution is the commissioning and use of interdisciplinary research, as this overcomes the divisions between disciplines within government departments through collaboration at an earlier stage in the policy process. Lowe and Phillipson (2006, p. 167) argue that interdisciplinarity emphasizes 'interaction and joint working, which brings the knowledge claims and conventions of different disciplines into a dialogue with each other, yielding new framings of research problems'. Whilst different modes of interdisciplinarity have emerged, their essence is a recognition that *interdisciplinary problems* are constituted relationally 'through dialogue or dissatisfaction with the problematics proffered by existing disciplines and institutions' (Barry et al., 2008, p. 30). Biosecurity neatly fits these emergent interdisciplinary approaches owing to the complex balance of attitudinal, behavioural, geospatial and epidemiological factors involved. But these approaches can also address the problem of communicating scientific knowledge about biosecurity. As this chapter has shown, social scientists have often assumed a 'bolt-on' or 'end-of-pipe' role, where they advise government on the implementation of policies which had been developed purely from natural science advice. In the words of the UK Commission on the Social Sciences (2003, p. 29):

[The role of] social sciences as a 'back-end fix' to the problems arising from new scientific developments ... can be parodied by 'we have invented this, now find a market for it' or 'we have invented this but it has a few unfortunate side effects. How do we get people to accept it?'

In response, some governments and research councils have begun funding major interdisciplinary research, particularly in the area of animal disease, agriculture and environmental change. In the UK, the roots of these initiatives stem from a series of agricultural policy crises, including bovine spongiform encephalopathy (BSE), the FMD epidemic and the opposition to genetically modified (GM) crops, which redirected attention towards public-interest science and reinforced arguments for joined-up approaches to rural policy. Together these changes reflected reorientations in the broader framing of public policy – from primary production to sustainable development, from a production-driven logic to one more oriented to the consumer, and from a sectoral to a territorial outlook in the management of rural areas – which in turn demanded an accompanying shift in the research base (Lowe and Phillipson, 2006, pp. 170–171). For biosecurity, this has led to research examining the risks of contracting E-coli 0157 (Strachan et al., 2011) and plant disease (Mills et al., 2011) that has combined scientific and 'lay' perceptions of disease with scientific understandings in order to build more effective disease management frameworks. For animal diseases like bTB, interdisciplinarity can help develop better epidemiological models of the disease by capturing farmers' own understandings of what counts as risky behaviour and recommendations for biosecurity practices that are based not just on what is effective, but what is practically achievable.

A key task for all these projects has been to incorporate different views into attempts to manage disease with the aim of identifying realistic solutions. It is not just enough for knowledge to be co-produced; it has to make a practical contribution to the lives of those living with biosecurity risks rather than simply being a pursuit of knowledge itself. However, interdisciplinarity is not without its challenges. First, interdisciplinary researchers face institutional difficulties of fitting into a regime of academic publishing and university structuring that does not favour boundary-crossing, is held in low esteem by mono-disciplinary colleagues and provides poor career progression opportunities (Bruce et al., 2004, p. 464). Interdisciplinarity can also require significant investment of time and resources to build up common vocabularies and analytical approaches between disciplines that are not used to each others' ways of working (Phillipson et al., 2009). Perhaps as a result, interdisciplinarity has also been criticized for promising to generate more accountable or innovative knowledge simply by bringing different disciplines together, regardless of the rigour or success of the outcomes (Barry et al., 2008).

Second, the involvement of stakeholders in determining research agendas can generate scepticism among those who feel that knowledge production and its application should be conceived of as logically distinct and separate. For these critics, 'the prospect of stakeholder engagement in knowledge production is typically viewed, at best, as a distraction and, at worst, as undermining scientific integrity' (Phillipson et al., 2012, p. 57). The extent to which altogether 'new' forms of knowledge can be created through interdisciplinary research is also contested (Bruce et al., 2004). Indeed, the challenge of satisfying all stakeholders seems particularly difficult. Whilst interdisciplinary studies of biosecurity have analysed farmers' own understandings and framings of risk in a bid to formulate better biosecurity policies, resolving their views with the attitudes of the wider public remains difficult. In New Zealand, for example, the use of a poison called 1080 to kill wild possums to prevent the spread of bTB has caused much public opposition out of concern for other wild birds and local water quality. In the UK, similar opposition has faced attempts to cull badgers as a way of controlling bTB (Enticott, 2001). Public opinion surveys frequently find a majority of the public against killing badgers or, indeed, other wildlife for biosecurity reasons, and a preference for other non-lethal management options (Dandy et al., 2012). Strategies of informing the public of scientific evidence have failed in the same way because they have not had the support of farmers. Not only have opposition groups become science-literate, but when presented with evidence about disease biosecurity policies they are also able to identify its uncertainties and assumptions (Defra, 2006). But in many cases, the acceptance of biosecurity appears to be related to sets of deep-seated moral philosophies of nature (Buller, 2008) that are not easily changed by scientific evidence. In short, interdisciplinarity offers no easy way of drawing together disparate and contrasting views on the management of animals, disease and the countryside.

The final challenge to interdisciplinary research is ensuring that findings are incorporated into policy making. Ethnographic studies of biosecurity policy making in the UK government reveal how biosecurity policy suffers from severe budgetary pressures, the constant threat of disease outbreaks and frequent organizational restructuring (Wilkinson, 2011). Together, these lead to a 'firefighting' approach rather than the longer-term strategic planning that is needed for a broad range of views to be incorporated into decision making. Within government departments responsible for biosecurity there is a very rapid staff turnover and consequent corporate memory loss that leads to much expertise that has built up among key policy makers being lost when they depart. As officials are often moved between jobs regularly they have little time to become 'experts' in their field and feel they are condemned to a treadmill of keeping up with new developments in their field and understanding both political and scientific issues relevant to their policy areas. Similarly, separate teams of officials are responsible for either policy making or evidence gathering, meaning that evidence specialists must battle to have their advice accepted, and policy makers may find that their demands for evidence cannot be met because research funding is not aligned with political priorities. Moreover, within the evidence teams, social, natural and veterinary scientists must compete for the same, increasingly scarce, resources to fund their research. Short-term, mono-disciplinary projects are more likely to receive support than large, multi-factorial interdisciplinary ones. Biosecurity policy officials therefore come to rely on a discourse of 'bureaucratic fatalism' to explain lack of progress or innovation in a particular area; the insistence upon following established procedures provides a rationale for inaction that is less risky for the individual bureaucrat than pursuing a radical policy that may fail (Merton, 1957; Crozier, 1964). Ironically, it is disease outbreaks that challenge the status quo and enable innovative ways of decision making to emerge (for example, borrowing military-style protocols during the 2001 FMD outbreak), but these prove difficult to sustain when the crisis has passed (Wilkinson, 2011). When it comes to governing biosecurity, these pressures may account for the continued reliance upon objective science and evidence as information to reduce uncertainty about the world, rather than more ambiguous social science and interdisciplinary research.

Conclusion

In this chapter we have attempted to show the various ways in which biosecurity is understood and made sense of by different groups including policy makers, scientists, vets, farmers and the wider public. Whilst expertise about animal disease is usually located within veterinary science, we have shown how the primacy of veterinary science itself was constructed in relation to the demands of society and the state. But the idea that veterinary science provides a universal way of understanding biosecurity is also flawed: we have shown that different forms of knowledge and biosecurity practices exist within veterinary science, whilst farmers themselves possess their own knowledges of what biosecurity practices may or may not work based on their own experiences of living with animal disease. Interdisciplinary forms of knowledge co-production may therefore offer a better insight into biosecurity controversies in future. Not only can these draw on different forms of biosecurity knowledge, but they can also help to resolve issues of uncertainty that plague the application and acceptance of biosecurity practices. To be sure, in other areas of agriculture, similar ways of thinking have been accepted for many years. Henke (2000), for example, describes how farmers test out and make sense of complex situations using their own knowledges, and those from scientists and trusted advisors. Perhaps the real mystery about biosecurity knowledges is why these social and interdisciplinary understandings have been missing for so long. Whatever the reason, we suggest that it is unrealistic to expect a resolution to the consequences of animal disease without the adoption of more social and interdisciplinary approaches to biosecurity research.

References

Anderson, I. 2002. *Foot and Mouth Disease 2001: Lessons to be Learned Inquiry Report*. London: The Stationery Office.

Backett, K. 1992. Taboos and excesses: lay health moralities in middle class families. *Sociology of Health and Illness* 14(2), pp. 255–274.

Barry, A., Born, G. and Weszkalnys, G. 2008. Logics of interdisciplinarity. *Economy and Society* 37(1), pp. 20–49.

Berg, M. 1997. *Rationalizing Medical Work: Decision-Support Techniques and Medical Practices*. London: MIT Press.

Bickerstaff, K. and Simmons, P. 2004. The right tool for the job? Modeling, spatial relationships, and styles of scientific practice in the UK foot and mouth crisis. *Environment and Planning D: Society and Space* 22(3), pp. 393–412.

Brown, J. S. and Duguid, P. 1991. Organizational learning and communities-of-practice: toward a unified view of working, learning, and innovation. *Organization Science* 2(1), pp. 40–57.

Bruce, A., Lyall, C., Tait, J. and Williams, R. 2004. Interdisciplinary integration in Europe: the case of the Fifth Framework programme. *Futures* 36(4), pp. 457–470.

Buller, H. 2008. Safe from the wolf: biosecurity, biodiversity, and competing philosophies of nature. *Environment and Planning* A 40(7), pp. 1583–1597.

Campbell, I. D. and Lee, R. 2003. Carnage by computer: the blackboard economies of the 2001 foot and mouth epidemic. *Social and Legal Studies* 12(4), pp. 425–459.

Crozier, M. 1964. *The Bureaucratic Phenomenon*. Chicago: Chicago University Press.

Dandy, N., Ballantyne, S., Moseley, D., Gill, R., Quine, C. and van der Wal, R. 2012. Exploring beliefs behind support for and opposition to wildlife management methods: a qualitative study. *European Journal of Wildlife Research* 58(4), pp. 695–706.

Davison, C., Smith, G. D. and Frankel, S. 1991. Lay epidemiology and the prevention paradox: the implications of coronary candidacy for health education. *Sociology of Health and Illness* 13(1), pp. 1–19.

Defra. 2006. *Public Consultation on Controlling the Spread of Bovine Tuberculosis in Cattle in High Incidence Areas in England: Badger Culling. A Report on the Citizens' Panels*. London: Defra.

Dreyfus, H. L. and Dreyfus, S. E. 1986. *Mind over Machine: The Power of Human Intuition and Expertise in the Era of the Computer*. Oxford: Blackwell.

Enticott, G. 2001. Calculating nature: the case of badgers, bovine tuberculosis and cattle. *Journal of Rural Studies* 17(2), pp. 149–164.

—— 2008. The ecological paradox: social and natural consequences of the geographies of animal health promotion. *Transactions of the Institute of British Geographers* 33(4), pp. 433–446.

—— 2012. The local universality of veterinary expertise and the geography of animal disease. *Transactions of the Institute of British Geographers* 37(1), pp. 75–88.

Hacking, I. 1983. *Representing and Intervening*. Cambridge: Cambridge University Press.

Hardy, A. 2003. Professional advantage and public health: British veterinarians and state veterinary services, 1865–1939. *Twentieth Century British History* 14(1), pp. 1–23.

Heffernan, C., Thomson, K. and Nielsen, L. 2008. Livestock vaccine adoption among poor farmers in Bolivia: remembering innovation diffusion theory. *Vaccine* 26(19), pp. 2433–2442.

Henke, C. R. 2000. Making a place for science: the field trial. *Social Studies of Science* 30(4), pp. 483–511.

Independent Scientific Group (ISG) 2007. *Bovine Tuberculosis: The Scientific Evidence*. London: Defra.

Jasanoff, S. S. 1987. Contested boundaries in policy-relevant science. *Social Studies of Science* 17(2), pp. 195–230.

Krebs, J., Anderson, R., Clutton-Brock, T., Morrison, I., Young, D. et al. 1997. *Bovine Tuberculosis in Cattle and Badgers*. London: Ministry of Agriculture, Fisheries and Food.

Latour, B. 1988. *The Pasteurization of France*. Cambridge, MA: Havard University Press.

Lowe, P. and Phillipson, J. 2006. Reflexive interdisciplinary research: the making of a research programme on the rural economy and land use. *Journal of Agricultural Economics* 57(2), pp. 165–184.

Lowe, P., Clark, J., Seymour, S. and Ward, N. 1997. *Moralizing the Environment: Countryside Change, Farming and Pollution*. London: UCL Press.

Lupton, D. and Tulloch, J. 2002. 'Life would be pretty dull without risk': voluntary risk-taking and its pleasures. *Health, Risk and Society* 4(2), pp. 113–124.

Merton, R. 1957. *Social Theory and Social Structure*. Glencoe, IL: Free Press.

Mills, P., Dehnen-Schmutz, K., Ilbery, B., Jeger, M., Jones, G. et al. 2011. Integrating natural and social science perspectives on plant disease risk, management and policy formulation. *Philosophical Transactions of the Royal Society B: Biological Sciences* 366(1573), pp. 2035–2044.

Orr, J. E. 1996. *Talking About Machines: An Ethnography of a Modern Job*. London: Cornell University Press.

Perren, R. 1978. *The Meat Trade in Britain 1840–1914*. London: Routledge.

Phillipson, J., Lowe, P. and Bullock, J. M. 2009. Navigating the social sciences: interdisciplinarity and ecology. *Journal of Applied Ecology* 46(2), pp. 261–264.

—— 2012. Stakeholder engagement and knowledge exchange in environmental research. *Journal of Environmental Management* 95(1), pp. 56–65.

Science Advisory Council. 2006. *Increasing the Capacity and Uptake of Social Research in Defra*. London: Defra.

Strachan, N. J. C., Hunter, C. J., Jones, C. D., Wilson, R. S., Ethelberg, S. et al. 2011. The relationship between lay and technical views of Escherichia coli O157 risk. *Philosophical Transactions of the Royal Society B: Biological Sciences* 366(1573), pp. 1999–2009.

Timmermans, S. and Berg, M. 1997. Standardization in action: achieving local universality through medical protocols. *Social Studies of Science* 27(2), pp. 273–305.

UK Commission on the Social Sciences. 2003. *Great Expectations: The Social Sciences in Britain*. London: Commission on the Social Sciences.

Waage, J. K. and Mumford, J. D. 2008. Agricultural biosecurity. *Philosophical Transactions of the Royal Society B: Biological Sciences* 363(1492), pp. 863–876.

Waddington, K. 2004. To stamp out 'so terrible a malady': bovine tuberculosis and tuberculin testing in Britain, 1890–1939. *Medical History* 48(1), pp. 29–48.

—— 2006. *The Bovine Scourge: Meat, Tuberculosis and Public Health, 1850–1914*. Woodbridge: Boydell Press.

Ward, N., Donaldson, A. and Lowe, P. 2004. Policy framing and learning the lessons from the UK's foot and mouth disease crisis. *Environment and Planning C: Government and Policy* 22(2), pp. 291–306.

Wilkinson, K. 2011. Organised chaos: an interpretive approach to evidence-based policy making in Defra. *Political Studies* 59(4), pp. 959–977.

Woodroffe, R., Donnelly, C. A., Johnston, W. T., Bourne, F. J., Cheeseman, C. L. et al. 2005. Spatial association of Mycobacterium bovis infection in cattle and badgers (*Meles meles*). *Journal of Applied Ecology* 42(5), pp. 852–862.

Woods, A. 2004. *A Manufactured Plague: The History of Foot and Mouth Disease in Britain*. London: Earthscan.

—— 2007. The farm as clinic: veterinary expertise and the transformation of dairy farming, 1930–1950. *Studies in History and Philosophy of Science Part C* 38(2), pp. 462–487.

—— 2009. 'Partnership' in action: contagious abortion and the governance of livestock disease in Britain, 1885–1921. *Minerva* 47(2), pp. 195–216.

—— 2011. A historical synopsis of farm animal disease and public policy in twentieth century Britain. *Philosophical Transactions of the Royal Society B: Biological Sciences* 366(1573), pp. 1943–1954.

Worboys, M. 1991. Germ theories of disease and British veterinary medicine, 1860–1890. *Medical history* 35(3), pp. 308–327.

Wynne, B. 1988. Unruly technology: practical rules, impractical discourses and public understanding. *Social Studies of Science* 18(1), pp. 147–167.

—— 1992. Misunderstood misunderstanding: social identities and public uptake of science. *Public Understanding of Science* 1(3), pp. 281–304.

—— 1996. May the sheep safely graze? A reflexive view of the expert–lay knowledge divide. *Risk, Environment and Modernity: Towards a New Ecology*, pp. 44–83.

7

BIOSECURITY MANAGEMENT PRACTICES

Determining and delivering a response

John Mumford

The management of biosecurity includes both preventative measures taken against potential perceived risks, which are based on formal risk assessments, and responsive measures taken as a result of the detection of outbreaks of harmful agents that have successfully entered a country. Because biosecurity requirements have implications for international trade they are governed by international standards and agreements. For plant health biosecurity affecting both agriculture and the environment, international standards cover pest risk analysis and risk management (FAO, 2004), and eradication actions (FAO, 1998). For animal health there are international standards on the veterinary inspection of exported animals and animal products, and diagnostic tests and control responses for all the major diseases of livestock (OIE, 2012a). The World Trade Organization (WTO) Sanitary and Phytosanitary (SPS) Agreement[1] describes general principles for biosecurity regulations and a dispute settlement procedure. This chapter considers these different preventative and responsive management practices, including exclusion, quarantine, early detection, surveillance, eradication and control. It incorporates a consideration of risk profiling and the allocation of response resources using economic risk modelling. Finally it reviews the sometimes contentious issue of cost and responsibility sharing.

Prevention

Much of the emphasis in international trade is on prevention of movement that leads to entry and establishment of harmful organisms in new areas, through a series of international organizations and treaties (Convention on Biological Diversity (CBD), International Plant Protection Convention (IPPC) and World Organisation for Animal Health (OIE)). The OIE was established in 1924 and is the longest running of the organizations recognized by the World Trade Organization Sanitary and Phystosanitary Agreement. Plant health is much more recent, with the IPPC dating from 1952, and its standard-setting role began as recently as 1989. The environmental agreement defined through the CBD was not established until 1992. The approaches taken by the different conventions reflect their differing emphases and the nature of the organisms they cover. The IPPC has provided general standards covering a wide range of pest management and assessment issues because the number of organisms affecting plant health is very substantial, whereas the OIE has focused on very specific standards, such as diagnostic procedures, for the limited number of livestock diseases. The CBD deals with longer timeframes and promotes a precautionary approach to uncertainty, while the WTO-related conventions require scientific evidence to support measures that could restrict trade. Each convention accepts the principle of national sovereignty in managing resources.

Risk profiling and risk management

Importing countries are responsible for risk analysis related to the commodities they import (FAO, 2004). Risk is commonly assessed by commodity pathway in many countries, such as the USA and Australia, in which the full range of potential pest species that may be associated with a particular commodity from a particular source country are listed and categorized for potential to enter, establish, spread and cause harm in the importing country. In the EU it has been more common to assess risks associated with individual species, regardless of their various pathways of entry, although both the European Food Safety Authority (EFSA) and the European and Mediterranean Plant Protection Organisation (EPPO) are increasingly looking at pathway risks. Risk assessments may also be initiated because of changes in phytosanitary policy or in response to changes in treatment measures or performance. The rationale for pathway assessments is that management can be applied at points along the specified routes from the field to the port of entry and this provides an effective and efficient basis for preventative risk management. Species assessments provide more information that would be relevant to the response in the importing area, such as surveillance planning and emergency post-entry response actions. Species assessments would provide information relevant to preventative management of the species on various pathways, but would not necessarily affect the management of a wider set of possible pests that may also be associated with any particular pathway.

The choice of risk assessment will be affected by the regulatory regime. Regimes that are aimed at excluding all potential exotic pests would imply pathway analyses, which should exhaustively catalogue all potential pests on the pathway. This is a difficult task because it is not always clear which organisms can enter a particular pathway, nor how great an impact an organism may have in a new and potentially more favourable environment. Furthermore, the organism may not be directly associated with the particular commodity in question. Regimes that name particular unwanted organisms, such as the European Council Directive 2000/29/EC Harmful Organism List, imply species assessments directed specifically at the listed harmful organisms. Pathway regimes, such as in the USA, reflect greater caution, since they may extend to many species of marginal risk among the complete pathway list, while it can also be hard to determine that the last potential species of concern has been identified in an assessment. Species regimes, as used in Europe, may be open to criticism for failing to include serious pests until they are added to the harmful organisms list.

Various risk assessment schemes are in operation in importing countries. For agricultural and environmental plant health assessments, many are carried out within the broad specifications outlined by the International Standard for Phytosanitary Measures no. 11 (FAO, 2004). The European and Mediterranean Plant Protection Organisation has a pest risk analysis scheme (EPPO, 2011) that can address either species or pathways, which establishes likelihoods and conse-quences of risk components for entry, establishment, spread and impact (Schrader et al., 2012). The European Food Safety Authority (EFSA) has an oversight role in pest risk analysis for member states in the European Union (EFSA, 2010), particularly focused on specific harmful organisms. In Great Britain, the risks from non-native species affecting the environment are assessed in a species-focused scheme closely related to that used by EPPO (Mumford et al., 2010). For animal health, risks have been identified at an international level and specific inspection, diagnostic and treatment protocols have been determined by the OIE (2012a).

The international standard ISPM 11 describes the transition from risk assess-ment to risk management (FAO, 2004). In many countries, such as the USA, the risk assessment and risk management stages of pest risk analysis are separated to distinguish the objective process of assessment and the application of value judgments and choice in management. Risk management consists of identifying options to respond to a perceived risk, evaluating performance and choosing the most appropriate actions. Uncertainty in the assessment should be taken into account in deciding how to respond with management actions. The aim of risk management is to achieve an acceptable level of risk, not necessarily to reduce risk to zero. This acceptable level of risk can be expressed in relation to other phytosanitary requirements in the importing country or another country facing the risk, measured in relation to estimated losses, or expressed on a ranked scale of tolerance. Additional management is not justified if the risk is already accepted, for example where a pest has become well established.

Risk management can be specified in terms of a standard for the acceptable level of a risk, or it could be specified as the application of a specific risk mitigation measure with an implied performance intended to meet such a standard. If the acceptable level of risk is specified it could be achieved by various different measures, and the outcome would be monitored as a frequency of an event, such as an outbreak, occurring within a defined area. If the measure is specified then the outcome would need some audit to demonstrate that the measure was performed, and some verification that performance was as expected. This raises the issue of equivalence when different measures are available to achieve control (FAO, 2005). The outcome of both these approaches would be affected by the extent of the challenge from the pests along the pathway, which is often difficult to know. The commodity supply chain often involves products grown and shipped from many different sources and channels, and even from a range of countries as the harvesting season progresses. The level of pests affecting the produce, and the performance of a range of control practices, are likely to be different in the different circumstances, making it difficult to specify the performance of control.

Quarantine and surveillance

Official quarantine management measures imposed as a result of a risk assessment should be efficient and practical, have minimal impact and not be additional to existing measures that are known to be effective. Alternative measures with the same effect should be accepted, and different measures should not be required on different pathways with the same risks (FAO, 2005). The equivalence of different management measures has been a subject of considerable debate over the years. Within trading agreements countries are required to accept equivalent measures, but the performance of measures is always probabilistic, so while two measures may have the same mean performance, the tails of their performance distributions may differ. Inspection at entry by the importing authorities provides a measure of the overall standard of management within a supply chain pathway, but it may not be able to distinguish whether an acceptable standard has been achieved because of low pest challenge or due to effective control measures.

Despite commercial quality management and regulatory requirements for quarantine measures, importing countries maintain surveillance systems to detect outbreaks or presence of exotic pests that occur beyond entry points for plant (FAO, 2011b) and animal (OIE, 2012a) health. Harmful organisms may be detected through active, planned surveillance operations or as a result of passive, informal observation. Planned surveillance may be the result of priorities established at a national or regional level, for example for key exotic pests of major crops, or because of a history of interceptions at entry ports or borders, or because of outbreaks reported in neighbouring countries or significant trading partners. Surveillance is inevitably of variable efficiency across a range of species, because the population densities of the organisms differ and the efficiency of trapping, observation and so on, is different for each method and species. General surveillance is not directed at particular organisms and can therefore not have any statistical basis related to a specific risk. General surveillance is also more difficult to assign, as a specific cost item, to particular pest control programmes.

Once an exotic harmful organism has been detected and taxonomically verified there must be an appropriate reporting process, both within the country to growers and other stakeholders, and internationally to ensure that other countries can take appropriate actions. Reports of new pests are notified to the IPPC for plant health, and to the OIE for animal health notification. Notifications are available through the respective websites. While notifications are essential to indicate that pests are newly present in a country, they do not reveal the level or efficiency of surveillance.

Surveillance may determine the presence of a new organism in a country, but the pest may be established in only a limited area. To retain an area designated as pest-free there would need to be a more detailed specific survey of sufficient intensity to demonstrate lack of presence, and evidence of measures to prevent entry and establishment from the infested portion of the country (FAO, 2011a). The choice of what area to attempt to delimit as pest-free would depend on an appraisal of the value of maintaining any trade advantages from the pest-free status.

In the EU, harmful organisms that are considered to be significant plant health risks are listed as an annex to the Plant Health Directive (European Commission, 2000) after agreement by competent authorities across the member states. Approximately 250 species are currently designated as Harmful Organisms, but surveillance is only compulsory for member states in declared emergencies, as part of official control measures, or in declared Protected Zones (FCEC, 2010). In the USA priorities for plant pest surveillance are set by the US Animal and Plant Health Inspection Service (APHIS) through the Cooperative Agricultural Pest Survey (CAPS) programme with the states (USDA, 2005). The CAPS programme maintains surveillance for approximately 400 plant pest species determined by ranking risks of economic and environmental impacts across the individual states.

The EU has specified statistically-based criteria for a reduced inspection for plant health pests based on the history of consignment numbers and past inspection results (European Commission, 2004). Consignments of commodities within the scheme may be subject to reduced checks on entry, expressed as a lower proportion of consignments inspected, based on the volume imported, the number of consignments on which harmful organisms have been intercepted, the estimated mobility of the harmful organisms associated with the commodity pathway and any other factor that might affect the risk. In this way, some risks can be continually updated and responsive management made more efficient and proportionate to the demonstrated risk.

Changes in trade patterns can result in rapidly changing priorities for pest detection and management. During the past two decades the massive increase in international trade in agricultural commodities has increased the risk of new pests, and the much greater share of EU and US imports originating from Asia has shifted the balance of risks away from species introduced from the more traditional transatlantic trade introductions (Waage and Mumford, 2008). In 2009, Thailand accounted for approximately 60 per cent of all interceptions of harmful organisms in the EU (FCEC, 2010). Overall, there has been a sharp rise in plant health interceptions notified in the EU since 2005, with approximately three times the number of interceptions in 2005–2009 compared to 2000–2004 (FCEC, 2010).

Eradication and pest management

The international standard dealing with plant health eradication measures (FAO, 1998) encourages the development of contingency plans so that actions can be taken at the earliest practical stage once an outbreak of a new harmful organism is detected. This planning process has been well developed in Australia in the form of Emergency Plant Pest Response Deeds (Plant Health Australia, 2011). In Australia, government and industry organizations have established formal con-tracts which specify the pests of concern to all parties, the steps that will be taken to prevent them and the cost- and responsibility-sharing arrangements to be put in place in the event an outbreak occurs. These agreements are intended to ensure that the relevant stakeholders have a say in decision making, from setting initial priorities through to how management actions are funded. The scheme for the deeds establishes categories of shared payments, from wholly government funding in the case of pest outbreaks that almost entirely affect a public good, to industry funding for pests that affect purely commercial interests with little public loss.

Decisions to eradicate a harmful organism that has entered will depend on the extent of the initial outbreak and the expectation that eradication measures can be effective. An analysis of the pathway for entry and establishment would be essential to ensure that the likelihood of further incursions can be reduced or eliminated. The NZ MAF (2002) has presented a useful general scheme for responding to pest outbreaks, from the immediate investigation of the situation in the field and its source, through coordinated operations and communications, to an eventual orderly closure of operations and learning exercise. The latter is particularly important as unwanted introductions are often repeated along the same pathway unless further preventative steps are taken. Sunley et al. (2012) describe a new decision support scheme to help set priorities for emergency management for post-entry decisions so that they can be made rapidly and with consistent comparisons.

Weighing the costs, benefits, risks and the capacity to respond

In principle, all harmful organisms, regardless of the type of organism and the nature of the impact, pose risks that should be managed proportionately. However, the nature of the impact and social imperatives, and the cost and practicality of management, will have important effects on the way management is conducted in practice, particularly for the question of whether to eradicate or suppress an invading organism (Fraser et al., 2006). Figure 7.1 indicates three general classes of harmful organism impacts, affecting human or animal disease, crop pests and organisms affecting the natural environment. Some organisms have immediate impacts of high value, such as human health or trade restrictions on animal products. Others, such as crop pests, have impacts that increase proportionally as they spread and impose yield losses and control costs across the sector. Many organisms that are harmful to the natural environment may have long delays in

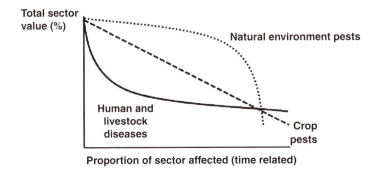

Figure 7.1 Relative time responses for impacts from different classes of harmful organisms
Source: After Waage et al. 2005

their effects, because early stages of infestation and spread have minimal impacts on the landscape or amenity value of the natural environment, and it is only once a large proportion of the host population is affected that the public perception of loss develops. These three impact curves demonstrate the rationale for high public investment in biosecurity for human and animal health and the relatively low spending on environmental biosecurity (Waage and Mumford, 2008).

The heart of a decision to eradicate would be a cost–benefit analysis, or business case, for action (Kehlenbeck et al., 2012; Mumford, 2007; Mumford, 2005). Temporal and spatial boundaries need to be determined for such an analysis. The spatial boundaries may be determined ecologically, such as the contiguous area of climatic and host suitability, or may be based on political boundaries, such as national borders. The temporal limits for an analysis can be more difficult to justify. While control actions for eradication may end within a fairly short period, there are likely to be ongoing preventative costs, and it may be argued that the benefits of eradication could last for a very long time. However, the uncertainty associated with realizing benefits long into the future, either because of reinvasion or other ecological or management changes, means that unless the benefits outweigh costs over periods of 10–20 years it is unlikely that eradication would be justifiable. Exceptions could include long-term conservation benefits, such as the elimination of rats and other non-native predators on sub-Antarctic islands.

In an analysis of US losses from introduced species, approximately a quarter of the lost value is in categories that cover 'environmental' effects, rather than affecting crops or domestic animals (Pimentel et al., 1999). The time dimension for environmental pests highlights the importance of the choice of discount rates in setting priorities (Mumford, 2002). Eradication is always easier and more efficient at the early stages of an invasion, but if the economic impact that justifies eradication is too distant in the future, decisions may not be made soon enough to be effective in eradicating an outbreak. Lower discount rates (United Kingdom Treasury, 2011) for long-term environmental losses could be used as part of the justification for earlier action, by increasing the estimated present value of long-term losses.

Costs and benefits must be compared to a baseline scenario representing the situation prior to the arrival of the pest. The specific control measures to be taken must be specified so that appropriate estimates of uptake, costs and performance can be made. While eradication may be quite fast, in some cases it can take many years and the time flow of costs and benefits would need to be considered for long-term programmes. Like risk assessments, cost–benefit analyses are probabilistic, so an estimate of the distribution of outcomes, or a sensitivity analysis, should be determined. Kehlenbeck et al. (2012) give some specific examples of cost–benefit estimates for outbreaks of Asian long-horned beetle (*Anoplophora glabripennis*) and Western corn rootworm (*Diabrotica virgifera*).

The response to harmful organisms should reflect both the risk they pose and the practical opportunities to manage those risks. It should be noted that even where non-native species are judged to pose low-level risks, these risks are additional burdens on the environment and the economy. Measures may be taken to protect against relatively low-risk species if the measures available are inexpensive and easily carried out. For example, while red-eared terrapins pose a relatively minor risk in the UK they have been banned from sale. On the other hand, difficult, expensive or ineffective measures may not be implemented even in cases where high impacts have been assessed, such as for rhododendrons in the UK (but see Taylor et al., Chapter 14, pages 218–219).

However, the required information to determine appropriate management responses is not always available. The recent outbreaks of Schmallenberg virus, a disease affecting sheep, cows and goats, in north-west Europe exemplifies the problems associated with dealing with unknown organisms (OIE, 2012b). Early efforts have demonstrated causality of virus and disease symptoms, and have identified related viruses to help establish diagnostics and possible epidemiology, and likely transmission mechanisms have been hypothesized. Transmission is currently assumed to be by small biting flies with some long-range dispersal by wind bringing the disease into the UK from mainland Europe. Animal disease outbreaks of this type can be brought under control by continual monitoring and destruction of infected herds, particularly during the winter when vectors are less active. However, wind-borne introduction could continue, so eradication would require extensive monitoring and infected herd destruction across Europe to be effective.

Eradication does not just involve actions by government and industry in all cases. Cooperative programmes in southern Brazil have attempted to eliminate the codling moth (*Cydia pomonella*), a European pest of apples, through pesticide treatments in large commercial orchards and by destruction of host trees in private gardens in affected communities (Kovaleski and Mumford, 2007). While government agricultural officials supervised the location and removal of the host trees, the orchard industry provided the labour and the replacement trees in an effort to share the responsibility for control.

From eradication to pest management

It is not always practical to eradicate an introduced pest, but a continued effort to slow its spread may be beneficial. An example of a long-running campaign to limit spread is the Gypsy moth (*Lymantria dispar*) in the USA (USDA, 2011). Gypsy moth was introduced in 1869 in New England, and an early effort was made to eradicate the pest in Massachusetts during the 1890s, which failed through lack of resources at the time. Subsequent efforts have focused on slowing the spread. The USDA estimates that the measures taken to reduce spread have slowed the rate of advance of the pest by approximately 16 km per year, which over a 20-year period is estimated to protect more than 61 million hectares of forest with a 20-year net present value (in 2007) of between US$184 and $348 million. While a campaign to slow the spreading impact of a pest may seem second best, once eradication is no longer feasible, it may be the only rational alternative. Also, while slow-the-spread campaigns for gypsy moth have run for over a century in the north-eastern USA, their economic appraisal can be done on a rolling short- to mid-term timeframe to ensure that efficient operations are maintained. In the case of European gypsy moth in the USA it is a practical action, because the spread is relatively slow because female moths do not fly. A new threat now looms in the Pacific Northwest, where Asian gypsy moths (*Lymantria dispar*) have been intercepted. The Asian gypsy moth females can fly, up to 30 km, which makes practical measures to reduce spread much more difficult if the initial efforts to eradicate primary outbreaks do not succeed. Where pests have the ability to spread fairly rapidly the efforts must be focused on the pest itself, whereas pests with more limited ability to spread can be addressed on a location-based approach.

The mechanism and rate of spread are key elements in the viability of an eradication or containment programme. The Western corn rootworm is an American pest of maize that has entered several European countries in the past 20 years (Carrasco et al., 2010). The pest lays its eggs in maize fields and is closely associated with continuous cultivation of maize. Eradication and containment involves crop rotation and insecticide treatment in outbreak areas and in a buffer zone surrounding the areas of known infestation to limit the spread. An immediate question for such operations is the width of the buffer zones needed to give effective control. Carrasco et al. (2010) provide an example of a quantitative estimation of the proportion of adults that would escape a combined focus zone (rotations limited to maize grown once in 3 years) and a safety zone (maize grown once in 2 years). By extending the intense control of a focus zone to 5 km from known rootworm-infested fields and the safety zone to 50 km it was estimated that less than 10 per cent and 1 per cent of adults would escape beyond the respective zones. This then draws attention to possible satellite outbreaks caused by longer range, possibly human-assisted, spread. While Carrasco et al. (2010) addressed the spatial dimensions of pest control, Breukers et al. (2008) carried out an analysis of potato brown rot control over various time scales, concluding that lower monitoring frequency would increase long-term costs. Less frequent monitoring means that outbreaks, when discovered, are larger and so more expensive to manage. The long-term costs and efficacy of monitoring and managing outbreaks must be set against efforts to prevent entry or establishment of exotic pests. In the case of gypsy moth, preventing entry is the main option, but for corn rootworm rotational practices can greatly reduce the chance of establishment and may provide a lower-cost option to reduce long-term impacts. Apart from the costs involved, different management options for preventing entry or establishment would place the burden of effort on different stakeholder groups.

Cost sharing and resource allocation

In Britain, as in many other countries, there is growing concern about the costs of animal and plant health measures to government, and there are moves to share some of these costs with a wider group of stakeholders. Important issues in any move to cost sharing include the need to provide some realistic estimate of the likely mid- to long-term cost distribution that stakeholders might be expected to share, the level of responsibility they would take, the capacity they would need to meet those responsibilities, and how stakeholders would contribute to decision making. Waage et al. (2007) described various limiting mechanisms for such schemes, in the context of plant health. Gosling et al. (2012) have estimated the distributions of outbreak costs that would be borne by government, under present arrangements, for a range of known animal diseases, as well as for an unknown disease. The estimates are based on the expert opinions of vets and economists. The estimated average annual outbreak cost for all the named, known animal diseases comes to £18 million, with the middle 50 per cent of the distribution ranging from about £15 million to £24 million. For the group of known diseases, 42 per cent of the uncertainty that accounts for this range of estimates comes from a single disease, foot-and-mouth disease (FMD). The inclusion of an unknown disease, however, dramatically changes the stakes. The estimated average annual outbreak cost for an unknown animal disease would be £15 million, almost as much as all the known diseases combined, with a very long upper tail of potential costs. The unknown disease would account for 98 per cent of the uncertainty in the estimated overall distribution of outbreak costs. Stakeholders would face a very substantial risk if they sign up to cost sharing that includes presently unknown diseases, or for a disease like FMD that has a very wide range of credible outbreak costs. As a result, governments may continue to bear the risk for where there is greater uncertainty, while arranging co-responsibility schemes for more limited risks.

Ideally the polluter, or introducer of pests, should pay for biosecurity (Waage and Mumford, 2008). However, it is difficult to organize biosecurity in such a way. Outbreaks may result from single introductions, or from a series of related or unrelated introductions, and continue through natural spread. Some of the ensuing spread may be the responsibility of inadequate initial responses (Mumford, 2011). It is therefore often very difficult to define individual responsibility for an outbreak. In any event, preventative management must be taken before an outbreak, which may not happen if prevention is successful, so the only practical bearers of the costs are the beneficiaries, either individually or through the state acting for the public good. Fines may help to cover some costs retrospectively if the cause of an outbreak can be proven, but they still only pay for failures, not success.

Insurance schemes have been suggested as one means of spreading costs and responsibilities for outbreaks of harmful organisms (Waage et al., 2007). While insurance would help to provide funding for control actions against the pests, it would also help to encourage good agricultural practice that would reduce the likelihood of outbreaks in the first place. An insurance scheme in the Netherlands to compensate for mandated destruction of potato crops infected with several important potato diseases has had high uptake from growers and requires, as a condition of subscription, that surface water is not used for irrigation. This has greatly reduced the frequency of outbreaks.

A critical issue in any response to an outbreak, particularly for unanticipated species, is the financial capacity to deal with containment and eradication. This has been evident in the case of the pinewood nematode (*Bursaphelenchus xylophilus*) in Europe. Since 1999 over €40 million has been spent in Portugal, with the removal and destruction of more than 5.5 million coniferous trees (FCEC, 2010). This effort has been supported by over €20 million from the EU solidarity fund for plant health.

Regional support for the management of plant pest outbreaks has been increasing in recent years in the EU. Prior to 2005, EU plant health solidarity payments to support national control efforts of regional importance generally ran at less than €1 million per year, but climbed to over €14 million in 2009 (FCEC, 2010), particularly because of the pinewood nematode in Portugal. This is symptomatic of a worldwide increase in outbreaks of tree pests, also including the Asian longhorned beetle and the emerald ash borer (*Agrilus planipennis*). The wood-boring pests have been moved around the world in wood-packaging material and pallets, and have led to several eradication programmes, including a successful Asian longhorned beetle eradication in the USA ending in 2005 and an unsuccessful attempt to eradicate emerald ash borer in the USA since 2002. The international reaction to the wood packaging problem led to the IPPC standard ISPM 15 in 2002 (FAO, 2009) and the implementation of certified wood treatment for all pallets and other packaging material in international trade.

The regional support programme for plant health in the EU (FCEC, 2010) has limitations because each country can only draw upon the fund for one emergency at a time, to prevent the fund becoming too open-ended. It does not currently support control of natural spread of pests, so countries must demonstrate that a pest has entered the country from outside the EU on consignments of imported products. The EU fund only covers up to 50 per cent of the cost for authorized control paid from public funds and it does not cover any of the commercial value of destroyed crops or lost income. These restrictions are necessary to maintain budget limits, but they can lead to artificial distortions of public and private responses where specific funding criteria are not met.

Cook et al. (2011) discuss the need for a more comprehensive economic framework that considers not only benefits and costs to the domestic producers protected by biosecurity measures imposed on imports, but also the welfare of consumers who may face higher costs and more limited choice as a result. Wider analysis such as this may have significant implications for the choice of organisms to be put on quarantine lists and for the way in which public policy determines what biosecurity should be funded.

Responsibility sharing: public vs private

Because biosecurity affects cross-border trade its regulation and implementation is generally considered to be an official function of national regulatory authorities operating for the public good. Phytosanitary certificates are issued by national authorities for exported goods, and inspections and mandated controls on imports are authorized by the national authorities of the receiving countries. However, some actions are delegated to private operators, such as the European Plant Passport scheme in many EU countries, or the industry-led control operations specified in Australia's emergency response deeds. Delegation may make some routine activities more efficient and may ensure more effective cooperative involvement from industry for emergency actions.

Conclusion

Harmful organisms may be assessed and managed at various stages as they move along the pathways on which they could enter and be established in a new country. While much of the regulatory emphasis is on planned preventative actions, this must be based on effective risk assessment processes to identify key points at which risk management is likely to be effective and efficient and on a clear assignment of responsibilities for risk management. Risk assessment and risk management have been based on pathways (covering a multitude of pests) or on particular species (across a range of pathways), but pathway-based analyses are becoming more prevalent. Pathway risk analysis can be more efficiently directed to management actions within the supply chain, where many aspects of quality and logistical management are already being applied. For unexpected arrivals, and in cases of natural spread, management is inevitably post-entry, but effective and efficient management again relies on clear responsibility for decisions and actions. Preventative management is primarily based on risk analysis, while responsive post-entry management of outbreaks may be based on cost–benefit analysis or, in some cases, on defined international obligations to take control actions to contain or eradicate particular organisms in the new environment. Agreed plans for co-responsibility between government and private stakeholders are becoming increasingly common in dealing with responsive management.

Note

1 www.worldtradelaw.net/uragreements/spsagreement.pdf.

References

Breukers, A., Mourits, M., Werf, W. v. d. and Lansink, A. O. (2008) Costs and benefits of controlling quarantine diseases: a bio-economic modeling approach. *Agricultural Economics*, 38: 137–149.

Carrasco, L. R., Harwood, T. D, Toepfer, S., MacLeod, A., Levay, N. et al. (2010) Dispersal kernels of the invasive alien Western Corn Rootworm and the effectiveness of buffer zones in eradication programs in Europe. *Annals of Applied Biology*, 156: 63–77.

Cook, D. C., Carrasco, L., Paini, D. R. and Fraser, R. (2011) Estimating the social welfare effects of New Zealand apple imports. *Australian Journal of Agricultural and Resource Economics*, 55: 599–620.

European and Mediterranean Plant Protection Organisation (EPPO) (2011) Guidelines on pest risk analysis. PM5/3(5) EPPO, Paris. 44pp.

European Commission (2000) Council directive 2000/29/EC. *Official Journal of the European Union*. L 169/10.7.2000. 159pp.

—— (2004) Commission regulation (EC) no. 1756/2004. *Official Journal of the European Union*. L 313/6. 4pp.

European Food Safety Authority (EFSA) (2010) Guidance on a harmonised framework for pest risk assessment and the identification and evaluation of pest risk management options by EFSA. EFSA, Parma, Italy. 66pp. www.efsa.europa.eu/de/scdocs/doc/1495.pdf.

FAO (1998) Guidelines for pest eradication programmes. ISPM no. 9, FAO, Rome. 12pp.

—— (2004) Pest risk analysis for quarantine pests including analysis of environmental risks. ISPM no. 11 (rev.), FAO, Rome. 26pp.

—— (2005) Guidelines for the determination and recognition of equivalence of phytosanitary measures. ISPM no. 24, FAO, Rome. 11pp.

—— (2009) Regulation of wood packaging material in international trade. ISPM no. 15 (rev.), FAO, Rome. 17pp.

—— (2011a) Requirements for the establishment of pest free areas. ISPM no. 4, FAO, Rome. 10pp.

—— (2011b) Guidelines for surveillance. ISPM no. 6, FAO, Rome. 9pp.

FCEC (Food Chain Evaluation Consortium) (2010) Evaluation of the Community Plant Health Regime. DG SANCO, European Commission, Brussels. Report 386pp. plus annexes 314pp. http://ec.europa.eu/food/plant/strategy/docs/final_report_eval_en.pdf; http://ec.europa.eu/food/plant/strategy/docs/annexes_eval_en.pdf.

Fraser, R. W., Cook, D. C., Mumford, J. D., Wilby, A. and Waage, J. K. (2006) Managing outbreaks of invasive species: eradication vs suppression. *International Journal of Pest Management*, 52: 261–268.

Gosling, J. P., Hart, A., Mouat, D. C., Sabirovic, M., Scanlan, S. and Simmons, A. (2012) Quantifying experts' uncertainty about the future cost of exotic diseases. *Risk Analysis*, 32: 881–893.

Kehlenbeck, H., Cannon, R., Breukers, A., Battisti, A., Leach, A. et al. (2012) A protocol for analysing the costs and benefits of plant health control measures. *EPPO Bulletin*, 42: 81–88.

Kovaleski, A. and Mumford, J. D. (2007) Pulling out the evil by the root: the codling moth eradication program in Brazil. In Vreysen, M. J. B., Robinson, A. S. and Hendrichs, J. (eds) *Area-Wide Control of Insect Pests: From Research to Field Implementation*, Springer, Dordrecht, Netherlands, pp. 581–590.

Mumford, J. D. (2001) Environmental risk evaluation in quarantine decision making. In Anderson, K., McRae, C. and Wilson, D. (eds) *The Economics of Quarantine and the SPS Agreement*, Centre for International Economic Studies, Adelaide and AFFA Biosecurity Australia, Canberra, Australia, pp. 353–383. 414pp.

—— (2002) Economic issues related to quarantine in international trade. *European Review of Agricultural Economics*, 29: 329–348. http://erae.oupjournals.org/cgi/reprint/29/3/329.pdf.

—— (2005) Application of benefit/cost analysis to insect pest control in crops using the sterile insect technique. In Dyck, V. A., Hendrichs, J. and Robinson, A. S. (eds) *Sterile Insect Technique. Principles and Practice in Area-wide Integrated Pest Management*, Springer, Heidelberg, Germany, pp. 481–498.

—— (2007) Model frameworks for strategic economic management of invasive species. In Lansink, A.O. (ed.) *Economics of Plant Health*, Springer, Dordrecht, Netherlands, pp. 181–190. 202pp.

—— (2011) Compensation for quarantine breaches in plant health. *Journal of Consumer Protection and Food Safety*, 6 (suppl.): S49-S54. DOI:10.1007/s00003-011-0674-7.

Mumford, J. D., Booy, O., Baker, R. H. A., Rees, M., Copp, G. H. et al. (2010) Invasive species risk assessment in Great Britain. *Aspects of Applied Biology*, 104: 49–54.

NZ MAF (2002) Cost–benefit analysis of unwanted organism or pest response options. http://brkb.biosecurity.govt.nz/processes-and-procedures/brm/response-management/develop-business-case/complete-cost-benefit-analysis/cost-benefit-analysis-of-unwanted-pest-response-options.doc.

OIE (2012a) Terrestrial animal health code. OIE, Paris. www.oie.int/international-standard-setting/terrestrial-code/access-online.

—— (2012b) Schmallenberg virus. OIE Technical Factsheet, OIE, Paris. February. 4pp. www.oie.int/fileadmin/Home/eng/Our_scientific_expertise/docs/pdf/A_Schmallenberg_virus.pdf.

Pimentel, D., Lach, L., Zuniga, R. and Morrison, D. (1999) *Environmental and Economic Costs Associated with Non-indigenous Species in the United States*. College of Agriculture and Life Sciences, Cornell University, Ithaca, NY, USA. 18pp. www.news.cornell.edu/releases/Jan99/species_costs.html.

Plant Health Australia. (2011) Government and plant industry cost sharing deed in respect of emergency plant pest responses. Canberra, Australia. 105pp. www.planthealthaustralia.com.au/go/phau/epprd.

Schrader, G., MacLeod, A., Petter, F., Baker, R. H. A., Brunel, S. et al. (2012) Consistency in pest risk analysis – How can it be achieved and what are the benefits? *EPPO Bulletin*, 42: 3–12.

Sunley, R., Cannon, R., Eyre, D., Baker, R. H. A., Battisti, A. et al. (2012) A decision support scheme that generates contingency plans and prioritises action during pest outbreaks. *EPPO Bulletin*, 42: 89–92.

United Kingdom Treasury (2011) *The Green Book*. HM Treasury, London. 114pp.

USDA (2005) *The Cooperative Agricultural Pest Survey. USDA, APHIS Program Aid 1830*. USDA, Washington DC. 2pp. www.aphis.usda.gov/publications/plant_health/content/printable_version/pub_phcapsdetecting.pdf.

—— (2011) Gypsy moth. Plant Health website, USDA, APHIS, Washington DC. USA. www.aphis.usda.gov/plant_health/plant_pest_info/gypsy_moth/index.shtml.

Waage, J. K. and Mumford, J. D. (2008) Agricultural biosecurity. *Philosophical Transactions of the Royal Society B*, 363: 863–876.

Waage, J. K., Fraser, R. W., Mumford, J. D., Cook, D. C. and Wilby, A. (2005) *A New Agenda for Biosecurity*. Defra, London. 198pp. http://horizonscanning.defra.gov.uk/Default.aspx?menu=menu&module=Program0205&NavID=36.

Waage, J. K., Mumford, J. D., Leach, A. W., Knight, J. D. and Quinlan, M. M. (2007) *Responsibility and Cost-sharing in Quarantine Plant Health*. Department for Environment, Food and Rural Affairs (Defra), London. 126pp.

Part III

BIOSECURITY AND GEOPOLITICS

8

A NEOLIBERAL BIOSECURITY?

The WTO, free trade and the governance of plant health

Clive Potter

Introduction

It is widely acknowledged that international trade – and the movement of animals, feed products, plants and plant materials that this entails – is heavily implicated in the growing number of animal and plant disease outbreaks around the world (Waage and Mumford, 2008; MacLeod et al., 2010). History is littered with examples of disease epidemics that can be traced to the importation of infected materials. The Irish potato blight outbreak of the 1840s probably originated in Atlantic shipments of seed potatoes that carried the fungus *Phytophthora infestans*, while the introduction of Dutch elm disease (*Ophiostoma novo-ulmi*) into the UK in the 1970s has since been attributed to the arrival of a single consignment of infected elm logs from North America (Gibbs, 1978).

Despite this history, the ability of individual governments and jurisdictions to restrict trade in order to reduce and manage risk is increasingly in conflict with the overarching, neoliberal agenda of free trade. Scholars of neoliberalization disagree about the extent to which it is possible to speak of a unified neoliberal project (Castree, 2008; Larner, 2003), but most agree that the market opening achieved through successive World Trade Organization (WTO)-sponsored trade rounds of the last 40 years is one of its defining characteristics. The WTO is the most important rule-making centre in the world for the promotion of free trade, with a prior commitment to dismantling tariff barriers to trade and facilitating market opening on neoclassically framed social welfare grounds. Many of the most significant trade disputes to have been brought before its Dispute Settlement Panel in recent years have a biosecurity dimension, as attempts by governments to limit trade in order to reduce threats to human, animal and plant health have been met with the counter-claim that these represent disguised barriers to trade.

While international agreements and protocols are in place to govern these disputes, notably the WTO's Sanitary and Phytosanitary (SPS) Agreement and its attendant Dispute Settlement Procedure (DSP), the boundary between biosecurity and international trade is an increasingly contested one, both internationally and within the boundaries of jurisdictions like the European Union (EU) (Maye et al., 2011). In this chapter, we explore the nature of this conflict and the protocols, risk assessment tools and predictive knowledge practices that have been put in place in an attempt to resolve disputes, manage risks and depoliticize debate. The chapter concludes by reflecting on the deeper contestations of trade, production and consumption that are emerging as part of a more critical response to plant health threats specifically.

We begin by describing a series of trade disputes that illustrate the opposed nature of the trade and biosecurity agendas but also the uneven path different governments have followed as they attempt to reconcile a macroeconomic commitment to open borders with the demands of domestic producers, consumers and citizens to protect themselves, their livelihoods and natural environments against pest and disease invasions. Plant biosecurity presents a particularly complex set of challenges in these terms and is the subject of the remainder of this chapter. Here, a traditional focus on preventing diseases affecting commercially grown crops is giving way to growing concern about the broader, public good impact of invasive pathogens and pest invasions on the natural environment and ecosystem services (MacLeod et al., 2010). This shift in concern is focussing critical attention on the international trade in plants and plant products that is a major risk driver for the spread of disease, but the absence of a commercial lobby or a coherent institutional voice pressing for import restriction means that the WTO has so far been only indirectly involved.[1] Rather than high-profile trade disputes, the emphasis is on improving the border inspection and quarantining procedures that are supposed to foster 'biosecure trade pathways'. But recent critiques of the fitness for purpose of the SPS and the international protocols determining how risks are assessed and assigned suggest that these procedures and methodologies may be flawed. Meanwhile, within jurisdictions such as the EU there is growing awareness of the limitations of its own Plant Health Regime (PHR), a set of standards, guidelines and recommendations that has co-evolved with the single market project and the need to harmonize approaches in order to reduce barriers to intra-EU trade. As we explain, this particular institutionalized attempt to reconcile free trade with biosecurity is coming under growing pressure as trade volumes expand and the practical limitations of border inspections, plant passporting and quarantining procedures in an EU of 27 member states become ever more apparent. The chapter looks specifically at the current threat posed by disease pathogens and insect pests to trees and woodlands to draw attention, not only to the limitations of these essentially managerial forms of neoliberal governance, but also to emphasize the restricted way in which the tree and plant health debate appears currently to be framed. While some critics are anxious to shift the debate away from a need to manage and contain diseases once they are introduced in favour of preventing their entry in the first place, this remains difficult within the terms of a largely technocratic and expert-dominated discourse on plant health.

Reconciling free trade with biosecurity under the WTO

Although it acknowledges that global trade poses risks to human, animal and plant health, the WTO has always been vigilant in seeking to minimize any disruptions to the free movement of goods, services and people that may be put in place on biosecurity grounds. As tariff barriers to free trade have been steadily reduced under successive trade rounds, so attention has refocused on the ability of governments to restrict imports using non-tariff barrier justifications such as these. Under the WTO's SPS Agreement (WTO, 2005), members have the right to impose restrictions when these are considered necessary to protect human, animal and plant health, but they are specifically prevented from implementing measures that impose any 'unnecessary, arbitrary, scientifically unjustifiable or disguised restrictions on trade' (Articles 2.2 and 5.1). They are further required to demonstrate a meaningful risk, using officially sanctioned and scientifically grounded risk assessment methodologies and tools.

One of the central provisions of the SPS Agreement is that reference must be made to the international standards and risk assessments set down by relevant international bodies. These are the Codex Alimentarius (CA) for food safety, the World Organisation for Animal Health (OIE)[2] for animal health and the International Plant Protection Convention (IPPC) for plant health. Together, these organizations develop the protocols for risk assessment that are often so influential in settling disputes and constitute the 'rule intermediaries' (Majone, 2002), which play a key role in helping make sense of the increasingly technocratic world of international biosecurity. Governments that base their import requirements on CA, OIE or IPPC standards are deemed under international law to have fully complied with the SPS. They may impose standards that are higher than those agreed by these organizations, but must then be able to provide a 'scientific justification' supported by a risk assessment (Hilson and French, 2003).

This refinement of procedure and technique, often at the expense of larger, more adversarial debates about conflicts, trade-offs and purpose, is one of the hallmarks of neoliberal governance. As Higgins and Dibden (2011, p. 396), drawing on Barry's (2001) concept of the 'metrological regime', observe, WTO-favoured tools such as risk assessment attempt to supplant political controversy, 'by allowing biosecurity and trade liberalization to be assessed within a single calculative space'. The black boxing of risk assessment is a precarious achievement, however, and challenges are still possible. In the case of biosecurity, the technical apparatus of 'dispute settlement' and the formal risk assessment procedures and knowledge practices with which it is associated have not always been successful in preventing disputes breaking out in the public realm. Differing interpretations of what constitutes a 'meaningful risk', rooted in divergent national commitments to defending food security, export capacity, rural environments and ecosystem services, mean that there have been a series of biosecurity-related disputes referred to the DSP over recent years (see Maye et al., 2011, for a fuller exposition and comparison of these disputes than is given here).

Notable amongst these is the series of trade embargoes and disputes surrounding the export of beef products from countries affected by the pathogen bovine spongiform encephalopathy (BSE). Various countries responded to the discovery of BSE in cattle in the UK during early 1986, with the US, for example, banning the imports of live ruminants from the UK in 1989 and all beef imports from 1997. At this stage in the outbreak, the US adopted a strong precautionary stance, arguing that the risk to human health from consuming infected meat was unknown but could be significant. It was therefore deemed legitimate to enforce a ban on imports of beef and beef products. The OIE played an important role in brokering debate and in defusing further trade disputes by adapting its advice as scientific understanding of the disease and its pathology developed and animal husbandry practices improved. The publication of a risk assessment for BSE resulted in a categorization of countries by disease risk (negligible, controlled and undetermined) and furnished a case for granting the UK a 'controlled risk' status in 2007.

Australia, however, with no previous history of BSE (partly due to its much earlier ban on animal feed imports because of concerns about the sheep disease scrapie) maintained its precautionary stance but as a result came under heavy international pressure to reopen its domestic market. Trading partners such as the US, Canada, Japan and other EU member states, as well as the UK, lobbied for a relaxation of a measure seen as a disguised form of trade protection. Despite protests from producer groups and domestic consumers, the Australian government eventually conceded that the quarantine regulations should be relaxed, an influential internal risk assessment having concluded that current import restrictions were not justified by the risk. The decision was nevertheless controversial within Australia, critics invoking the case as proof of the difficulty of maintaining a precautionary stance on domestic public health grounds in the face of international pressure from the OIE, WTO and other trading nations to maintain open borders (Bambrick et al., 2004).

The technical apparatus of risk assessment is not itself always beyond dispute, however, as can be seen in the much reported conflict between Australia and New Zealand concerning the importation of apples (Higgins and Dibden, 2011). New Zealand is an important producer and exporter of apples but since 1921 has been unable to gain access to its neighbour's markets following an import ban introduced by the Australians to prevent the introduction of fire blight (*Erwinia amylovora*), a bacterial infection that had led to widespread damage to commercial apple and pear orchards in New Zealand (as well as in the US and parts of Europe). In 1986 the New Zealand government formally requested access to the Australian market but this request was met with an authoritative Import Risk Assessment (IRA) undertaken by Biosecurity Australia which concluded that New Zealand apples continued to represent a transmission pathway for the disease. Despite subsequent attempts to agree a 'Standard Operating Procedure' (SOP), which would specify new phytosanitary procedures to be followed by exporters under a managed market opening, New Zealand eventually initiated WTO dispute settlement procedures. New Zealand disputed the scientific validity of the IRA and argued that the import ban was, in fact, a disguised form of protection for Australia's domestic apple sector. Australia in response mentioned its special phytosanitary status as 'a geographically isolated island with a relatively short history of agricultural production' (quoted in Gruszczynski, 2010, p. 15) and argued that New Zealand was seeking to impose a uniform risk assessment that was insufficiently sensitive to the particularities of the case. As Higgins and Dibden (2011) show in their analysis, a dispute apparently framed in technical, metrological terms, disguised a much more politicized conflict between free trade and biosecurity, which neither side was able to acknowledge or open up for debate.

Governing plant health in a neoliberal world

Risks to plant health have been much less often implicated in international trade disputes compared to threats to human and animal health. This partly reflects long-standing international arrangements through which agencies such as the IPPC have been able to harmonize standards, approaches and knowledge bases and practices (MacLeod et al., 2010), but also the greater ability to contain outbreaks within territories affecting agricultural crops using chemical treatments, albeit often at great expense and at the risk of collateral damage. Yet conflicts between national interests in protecting plant health and international commitments to free trade look set to intensify in future, as attention shifts away from a traditional focus on the impacts of pathogens and pests on production and commercial interests, to the broader threats of emerging diseases to natural environments and public goods such as food security, ecosystem services and biodiversity.

Plant health has a long history, beginning with the earliest attempts to protect wheat from black stem rust (*Pucciniagraminis*; Ebbels, 2003). The emphasis on the wider environmental effects of disease invasions is a relatively recent development, following decades in which plant health could be regarded as a branch of crop protection, chiefly concerned with the prevention of diseases and pests affecting commercially grown crops, primarily agricultural but also in the forestry and horticultural sectors. A reframing of the disease threat to agricultural production in terms of food security is an important development, linking plant health to broader debates about how best to safeguard the productivity and output of national agricultures as public goods at a time of impending global supply/demand imbalances (Winter and Lobley, 2010). But awareness of the damaging impact of invasive pathogens on the natural environment arguably introduces an even more novel dimension of concern that has been little analysed by social scientists until now.

Since the 1990s, a stream of invasive pathogens damaging to trees, forests and native plant communities has been closely linked to the commercial movement of living plants and the trade in hardy shrubs and ornamentals (Anderson et al., 2004). Twice during the twentieth century, pandemics of Dutch elm disease have spread throughout North America, Europe and south-west Asia, killing elm trees (*Ulmus* spp.) in large numbers, with often significant consequences for biodiversity and landscapes. Both pandemics were driven by the international transportation of infected timber. The movement of non-native plants and plant material, a more recent development in trade, across biogeographical zones is thought to pose an even greater threat because plant communities that have previously been exposed only to micro-organisms and viruses with which they had co-evolved are now coming into more frequent contact with novel pathogens to which they have limited or no resistance. Pathogens, meanwhile, can establish quickly and spread because they have no natural enemies. According to Daszak et al. (2000), the introduction of alien pathogens has until recently been one of the most under-estimated causes of global anthropogenic environmental change. Jones and Baker (2004), for instance, estimate that, in the UK, 67 per cent of newly arrived pathogens are associated with wild or ornamental plants, posing a significant threat to the natural environment and horticultural heritage. In the US, where successive disease invasions over the past decade have devastated native tree populations, the federal Animal and Plant Health Inspection Service (APHIS) has identified 'plants for planting' as a major disease pathway requiring improved levels of surveillance (APHIS, 2010).

The horticultural trade in ornamental plants and hardy shrubs is heavily implicated in this problem, having developed a much expanded capacity over the last 10 years to export to long-distance markets (Dehnen-Schmutz et al., 2010). Structural changes in the industry, linked to growing consumer demand for ornamental plants and mature and semi-mature trees and shrubs, mean that this is now a global market. Its trade pathways have become major transmission routes for plant diseases and insect pests (Hulme, 2009). Protocols exist but many date from the 1950s when plant health was conceived in a different way, empha-sizing the need to draw up lists of already identified harmful organisms that impact on agricultural commodities and timber products as a first step towards anticipating and preventing invasions. It is significant that until recently, the EU's Plant Health Directive made no reference to pests and pathogens that do not threaten commercially valuable species (Hulme, 2009). A feature of the invasive pathogens now threatening the wider environment is that many of the organisms were unknown to science before they were identified, which limits the usefulness of favoured methodologies such as Pest Risk Assessment (PRA) in providing a justification for action. As Brasier (2008) points out, 'sudden oak death', Dutch elm disease and the proliferating range of phytophthora diseases affecting a range of tree species and native plant communities worldwide did not appear on any international lists before they were identified.

Moreover, it is estimated that only 7–10 per cent of all fungal species with the potential to become pathogenic have so far been described. Many dangerous pathogens thus do not currently have PRAs, and the difficulty of being able to predict a pathogen's host range (which may be different to the species infected in the country of source), means that it is difficult to identify the risk pathways themselves. This condition of ignorance is seen by some critics as a reason to exercise precaution and suggests that the threat can only be contained in the long term by severely limiting the movement of plant materials around the world. Brasier (2008, p. 13), for instance, argues that 'the obvious and most effective way to reduce the risks would be to limit the level of plant imports to the minimum necessary for subsequent propagation'. But to frame the problem in this way would have profound implications for the horticultural trade and the EU's concern to maintain open borders. It also implies a problematization of the relationship between plant biosecurity and trade that the WTO and many industry lobbyists would be anxious to resist. Certainly any reduction in the *volume* of trade is unlikely given the huge commercial stakes involved.

Reconciling biosecurity with market rule under the EU's single market

Rather, the emphasis continues to be placed on improving the *practices* of biosecurity – those border controls, quarantining procedures, monitoring, surveillance and management interventions that largely define this policy domain – together with the predictive and diagnostic knowledge tools that underpin them. The EU's Plant Health Regime offers a particularly vivid illustration of the inherent limitations of this ameliorative, technically focussed approach, with a growing tension emerging between the macroeconomic priorities of the EU's single market and its efforts to install a risk-sensitive policy for managing the disease threat to its native plant communities, habitats and landscapes. These tensions between market rule and biosecurity are revealed at the level of an individual member state like the UK, whose approach to biosecurity reflects its participation in these broader EU processes of governance and market opening.

Prior to joining the EU, the UK had its own approach to plant health sanctioned under the 1967 Plant Health Act, but as a trading nation it was an early advocate of the need for harmonized international standards and procedures in order to minimize disruptions to trade. Nevertheless, the completion of the single market, arguably the defining neoliberal project of European integration, required the adoption of a new and more monolithic model of plant biosecurity if plant health was to be safeguarded following the dismantling of internal border controls. The founding principle of the EU's Plant Health Regime was that any plant material should be able to move without hindrance within the EU once it has cleared an external border. Risk management measures were applied to trade pathways into the EU and a new approach to intra-EU trade adopted. Under a system of 'plant passporting', introduced in 1993, any producer (chiefly commercial nurseries) wishing to move plants and plant material within the EU must be issued with a phytosanitary certificate (a 'passport') by the plant health authorities in their jurisdiction following an inspection to verify that it poses no threat to plant health.

In the UK, such duties are performed by Defra's Plant Health and Seeds Inspectorate (PHSI), but its ability to prevent new introductions under this harmonized, single-market-compatible system is dependent on the reliability of plant passports issued in other member states. As a recent review of the PHR acknowledges, the system works well for those trades which have a long record of compliance, but it is philosophically and operationally flawed as a system for managing the risks posed by previously unknown invasive pathogens (CEC, 2010). Founded on a 'weakest link public good' principle, the effectiveness of the Regime overall can only be as good as that of the weakest member state. Yet inspection and quarantining standards vary widely across the EU, with some member states having much less rigorous procedures than others in their approach to inspection and diagnostic testing. Beyond this, the emphasis on visual inspections of random samples of material present in consignments is increasingly problematic given that growing numbers of pathogens may be present as largely invisible propagules, mycelia or spores in the roots, leaves or substrate of imported plants. The consequences for the UK have been a series of disease invasions in recent years due to biosecurity lapses on imports of infected ornamental plants from elsewhere in the EU.

One of the most serious of these was the arrival and spread of the pathogen *Phytophthora ramorum* (Pr) in the early 2000s. This outbreak has been traced to a single introduction, probably on infected nursery stock that was brought to the UK via the European nursery trade (Harwood et al., 2009). The biological origins of the pathogen have been traced to Southeast Asia, where it is thought to have entered international trade pathways on exotic horticultural plants of various sorts. It had previously been identified in the US in 2001 and the resulting epidemic of what would be called 'sudden oak death' has infected millions of tan oaks and has seriously depleted the coastal forests of California and south-west Oregon (Rizzo and Garboletto, 2003). In 2002, it was recognized under the PHR as one of the most significant quarantine pathogens within the EU's jurisdiction and measures were put in place to limit movements of this quarantined pathogen. A fungal pathogen with an ever widening host range, the disease spreads via long-lived spores that are produced in large numbers and then disseminated through the natural environment via multiple pathways, including rain-splash and windblown leaves, via water-courses, on the footwear of countryside visitors and through animal movements. From an original outbreak in Cornwall in Southwest England first identified in 2002, the disease has expanded throughout the Southwest, into southern Wales and western Scotland. In the UK, the pathogen was initially found in woodland wherever its principal British host species, *Rhododendron ponticum*, was present as an understorey shrub. Rhododendron (or other shrubs, such as magnolias, where absent) is usually the first site of infection, before the disease spreads to susceptible trees such as beech (*Fagus sylvatica*), ash (*Fraxinus* spp.), sweet chestnut (*Castanea satira*), Japanese larch (*Larix kaempferi*) and European larch (*L. decidua*). The resulting threat to the UK's semi-natural woodland, heathland and commercial forestry is now deemed to be significant, with Defra speaking of the potential for 'a landscape scale epidemic' (Defra, 2011).

Once formally identified as a disease risk, the reaction of the UK's plant health authorities[3] was in many respects exemplary within the terms of conventional biosecurity practice, with measures rapidly put in place to contain the outbreak and prevent further imports of infected material (Tomlinson et al., 2009). Annual surveys of nursery stock were initiated, with a policy of destroying all infected plants found within 2 km of where infections were discovered. Meanwhile, an emergency programme was established in Southwest England to inspect woodland gardens and semi-managed or unmanaged woodland, and to destroy areas of R. ponticum where infection is found. In 2009, following a full science and policy review conducted in consultation with stakeholders (Defra, 2008), a new programme was established with increased funding, with the goal of 'reducing the innoculum to epidemiologically insignificant levels' through a more extensive programme of R. ponticum clearance and better containment and eradication measures in infected gardens and nursery sites.

Despite this, the plant health authorities have faced a considerable challenge in containing a disease that is now established in the UK. The recent history of the outbreak illustrates the limited extent to which invasive pathogens like these can be contained once they have been introduced through trade, with the disease continuing to spread and expand its host range. There were over 1,000 reports of infected trees by June 2011, with infections increasing in number and geographical spread. Some 80 outbreaks of a symptomatically similar Phytophthora kernoviae (Pk) have also been been notified, chiefly in Southwest England. Containment of both Pr and Pk is difficult for a number of reasons. First, the disease system is biologically complex and hard to diagnose. While mortality rates for most trees are lower than for Dutch elm disease, infection periods are much longer and may be asymptomatic, suggesting that the pathogen can be present but undetected for long periods. Visually healthy plants in woodland gardens in the Southwest, for instance, may be producing very large volumes of spores that continue to infect other plants. Equally, as we have commented above, port inspections based purely on visual inspection may still be admitting infected material, particularly in the substrate of trees and ornamental plants. Second, and most significantly in terms of being able to anticipate and limit spread, the host range is very wide and appears to be expanding. Indeed, the pathogen has expanded its host range in ways that could not have been predicted at the outset of the epidemic, spreading to infect commercial forestry trees such as Japanese larch and heathland habitats.

In acknowledging that the disease is now endemic in the UK and that future management must focus on containment rather than eradication, Defra and the Forestry Commission are following a well rehearsed script in tree disease outbreak management that begins with an attempt to eradicate, moves through a containment stage and concludes with (more or less effective) adaptation to what is effectively an endemic disease (Potter et al., 2011). Nevertheless, set against a background of a proliferating set of threats to tree health in the UK and elsewhere in the EU (Grunwald et al., 2012), the episode appears to have raised awareness amongst biosecurity professionals and policymakers about the need for more preventative action. The recently published *Action Plan for Tree Health and Plant Biosecurity* (Defra, 2011, p. 6) includes commitments to 'strengthen import control activities and protocols' and 'facilitate greater international collaboration to ensure that new trades and other potential pathways are safe'. Significantly, while Defra says it will contribute to the reform of the EU's PHR, it is unspecific about how wide-ranging these reforms need to be and does not respond to the requests of some critics for a more root-and-branch review of the horticultural trade and the biosecurity risks it poses. While some individuals are willing to ask whether the benefits to domestic consumers in terms of the availability of a wide range of low-priced ornamental plants is worth the potential cost to natural environments, this is not how the debate is typically framed.

Interestingly, compared to the attention given to the environmental impacts of non-native species by environmental and other lobby groups, invasive pathogens attract much less attention and concern. Instead, a largely expert-centric debate conducted between regulatory scientists, plant health professionals and policy-makers continues to focus on the need to improve inspection procedures, diagnostic testing and predictive modelling. The broader trade dimension is largely absent in these discussions. This further reflects low levels of awareness and political engagement amongst stakeholders, including, crucially, the plant-buying and gardening public, for whom nursery purchases, plant collecting and garden visits remain blameless activities with few, if any, collateral consequences (Potter et al., 2011). Those agitating for a wider debate about plant and tree health seem to be the forest and plant pathologists whose research in the field has alerted them to the threat (Clive Brasier, for instance, is a leading forest pathologist with many years' experience as a regulatory scientist working for the UK government). Ironically, it is their very knowledge practices that seem to be contributing to the continued closure of debate.

Conclusion

The defence of borders to 'keep out' potentially destructive disease pathogens and pests is arguably one of the dominant narratives of biosecurity policy and practice. But border controls and import restrictions are inimical to the open market that supporters of free trade believe to be necessary on social welfare grounds. The WTO solution is to exceptionalize import bans or trade restrictions that are made for reasons of biosecurity, placing the burden of proof onto the initiating countries and jurisdictions. At the same time, it seeks through the SPS to standardize the ways risks are measured and assessed and hence to forestall larger debates about the increasingly problematic relationship between international trade and the spread of diseases which threaten human, animal and plant life. Standard operating procedures are to be followed and international (OIE, CA and IPPC) standards benchmarked whenever disputes arise. An increasingly elaborate technical apparatus of risk assessment has become the methodology of first resort for resolving conflict between WTO member states. Yet, as we have shown, this technostructure is not always effective in preventing high-profile trade disputes breaking out into the public realm. Seen in these terms, the threat to public goods, such as biodiversity and various ecosystem functions, from a proliferating range of tree pathogens and pests presents a wholly new set of challenges. Trade embargoes are unlikely, given the public good dimension and the lack of an organized lobby or institutional voice willing to make the case for restrictions to the horticultural trade. Nevertheless, there are the first stirrings of a more critical debate here which may yet connect the plant-buying and plant-collecting habits of large numbers of consumers to a growing threat to native environments and the limitations of a version of biosecurity that is co-constituted with, rather than set apart from, the liberalization of international trade.

Notes

1 WTO-arbitrated plant trade disputes within the EU tend to be less common than those concerning animal diseases due to the existence of the EU's Plant Health Regime, which has authority to settle disputes between member states and between member states and other countries.
2 Originally entitled the Office International des Epizooties (OIE) but today still known by its original acronym.
3 The Department for Food, Environment and Rural Affairs (Defra) and the Forestry Commission (through the work of its Plant Health Service) share responsibility for managing the disease in England, while the Scottish Government and the Welsh Assembly Government share responsibility with the Forestry Commission in their respective jurisdictions.

References

Anderson, K., Cunningham, A., Patel, N., Morales, F., Epstein, P. and Daszak, P. (2004) 'Emerging infectious diseases of plants: pollution, climate change and agrotechnology drivers', *Trends in Ecology and Evolution*, vol. 19: 535–544.

APHIS (2010) *APHIS Strategic Plan, 2010–2015*, United States Department of Agriculture Animal and Plant Health Inspection Service, Washington DC.

Bambrick, H., Broom, D. and Denniss, R. (2004) *Public Health Impacts of the Proposed Australia–United States Free Trade Agreement: Pharmaceuticals and Food Safety.* Submission to the Senate Select Committee on the Free Trade Agreement between Australia and the United States of America, 30 April.

Barry, A. (2001) *Political Machines: Governing a Technological Society*, Athlone Press, London.

Brasier, C. (2008) 'The biosecurity threat to the UK and global environment from international trade in plants', *Plant Pathology*, vol. 57: 792–808.

Castree, N. (2008) 'Neoliberalizing nature: process, effects and evaluations', *Environment and Planning A*, vol. 40: 153–173.

CEC (2010) *Evaluation of the Community Plant Health Regime, Final Report*, European Commission DG SANCO, Brussels.

Daszak, P., Cunningham, A. A. and Hyatt, A. D. (2000) 'Emerging infectious diseases of wildlife – threats to biodiversity and human health', *Science*, vol. 287: 443–449.

Defra (2008) *Consultation on Future Management of Risks from Phytophthora Ramorum and P. Kernoviae*, Defra, London.

—— (2011) *Action Plan for Tree Health and Plant Biosecurity*, Defra and Forestry Commission, London.

Dehnen-Schmutz, K., MacLeod, A., Reed, P. and Mills, P. (2010) 'The role of regulatory mechanisms for control of plant diseases and food security: case studies from potato production in Britain', *Food Security*, vol. 2: 233–245.

Ebbels, D. (2003) *Principles of Plant Health and Quarantine*, CABI, Wallingford..

Gibbs, J. N. (1978) 'Development of the Dutch elm disease epidemic in southern England, 1971–6', *Annals of Applied Biology*, 88: 219–228.

Grunwald, N. J., Garbelotto, M., Goss, E. M., Heungens, K. and Prospero, S. (2012) 'Emergence of the sudden oak death pathogen Phytophthora ramorum', *Trends in Microbiology*, vol. 20, no. 3: 131–138.

Gruszczynski, L. (2010) 'The standard of review in international SPS trade disputes: some new developments', Research paper 3, Institute of Legal Studies, Polish Academy of Sciences, Warsaw.

Harwood, T., Pautasso, M., Jeger, M. and Shaw, M. (2009) 'Epidemiological risk assessment using linked network and grid based modelling: *Phytophthora ramorum* and *Phytophthora kernoviae* in the UK', *Ecological Modelling*, vol. 220: 3353–3361.

Higgins, V. and Dibden, J. (2011) 'Biosecurity, trade liberalization and the (anti)politics of risk analysis: the Australia–New Zealand apples dispute', *Environment and Planning A*, vol. 43: 393–409.

Hilson, C. and French, D. (2003) 'Regulating GM products in the EU: risk, precaution and international trade', in M. Cardwell, M. Grossman and C. Rodgers (eds), *Agriculture and International Trade – Law, Policy and the WTO*, CABI Publishing, Wallingford.

Hulme, P. (2009) 'Trade, transport and trouble: managing invasive species pathways in an era of globalisation', *Journal of Applied Ecology*, vol. 46: 10–18.

Jones, D. and Baker, R. (2004) 'Introduction of non-native pathogens into Great Britain, 1970–2004', *Plant Pathology*, vol. 56: 211–222.

Larner, W. (2003) 'Neoliberalism?' *Environment and Planning D*, vol. 21: 509–512.

MacLeod, A., Pautasso, M., Jeger, M. J. and Haines-Young, R. (2010) 'Evolution of the international regulation of plant pests and challenges for future plant health', *Food Security*, vol. 2: 49–70.

Majone, G. (2002) 'What price safety? The precautionary principle and its policy implications', *Journal of Common Market Studies*, vol. 40: 89–109.

Maye, D., Dibden, J., Higgins, V. and Potter, C. (2011) 'Governing biosecurity in a neoliberal world: comparative perspectives from Australia and the United Kingdom', *Environment and Planning A*, vol. 43: xx.

Potter, C., Harwood, T., Knight, J. and Tomlinson, I. (2011) 'Learning from history, predicting the future: the UK Dutch elm disease outbreak in relation to contemporary tree disease threats', *Philosophical Transactions of the Royal Society*, vol. 366: 1966–1974.

Rizzo, D. and Garbolotto, M. (2003) 'Sudden oak death: endangering California and Oregon forest ecosystems', *Frontiers in Ecology and the Environment*, vol. 1: 198–204.

Tomlinson, I., Harwood, T., Potter, C. and Knight, J. (2009) *Review of Joint Inter-Departmental Emergency Programme to Contain and Eradicate Phytophthora Ramorum and P. Kernoviae in Great Britain*, Defra, London.

Waage, J. K. and Mumford, J. D. (2008) 'Agricultural biosecurity', *Philosophical Transactions of the Royal Society*, vol. 363: 863–876.

Winter, M. and Lobley, M. (eds) (2010) *What Is Land For? The Food, Fuel and Climate Change Debates*, Earthscan, London.

WTO (2005) *The WTO Agreement on the Application of Sanitary and Phytosanitary Measures*, Geneva, World Trade Organization.

9

VIRAL GEOPOLITICS
Biosecurity and global health governance

Alan Ingram

Introduction

This chapter examines tensions surrounding the development and reworking of global health governance in response to concerns about emerging infectious diseases since the 1980s. It focuses in particular on how tensions have emerged at the intersections between technologies of government – what, following Michel Foucault, may be termed apparatuses of security – that have been created in response to newly formed infectious disease epidemics and struggles over the international political economy.

The widespread adoption of the term 'global health governance' can be understood as a result of a convergence between struggles over globalisation and growing unease about emerging infectious diseases. The intensification and expansion of international trade and travel, combined with environmental change, population growth, urbanisation and shifts in farming practices, is generally understood to have transformed the ecological matrix within which humans, animals, plants and microbes co-exist and co-evolve. The intensified interactions and transactions associated with globalisation are commonly understood to have heightened the risk of disease emergence into human populations and its subsequent spread.

In response to these quantitative and qualitative shifts in epidemiological space and time, materialised through a series of infectious disease outbreaks and epidemics. Health bureaucrats, scientists, politicians, activists, and corporate and military entities have collaborated and struggled over the creation of new organisations, networks and strategies, fostering the emergence of the field of global health (Lakoff and Collier, 2008).

In a lecture course given at the Collège de France in the late 1970s, Michel Foucault (Foucault, 2007) described the consolidation of such clusters of institutions, rationalities, tactics and technologies in response to crisis or emergency situations as the formation of apparatuses or mechanisms (words that provide an approximation to the word *dispositif* that Foucault used) of security. In an interview given around the same time, Foucault elaborated further on what he meant by this term:

What I'm trying to pick out with this term is, firstly, a thoroughly heterogeneous ensemble consisting of discourses, institutions, architectural forms, regulatory decisions, laws, administrative measures, scientific statements, philosophical, moral and philanthropic propositions – in short, the said as much as the unsaid. Such are the elements of the apparatus. The apparatus itself is the system of relations that can be established between these elements.

(Foucault, 1980, p. 194)

Crucially, in modern societies undergoing an increased scope, pace and intensity of circulation, apparatuses of security are characteristically concerned with the management of risks and dangers, seeking to forestall problems and to check them when they appear to be running out of control. Rather than the sovereign's command or the magistrate's prohibition, problems are to be managed in terms of scientific and economic expertise. In the context of governmentality (Foucault's term for the calculated, reflexive arts of government that emerged in Western societies), problems of government are to be understood less as political challenges and more as technical ones.

As the international relations theorist Stefan Elbe (Elbe, 2009) has argued, over the course of the last two decades, global health has become a key site for the *governmentalisation* of world politics, with the installation of rationalities and technologies for the control of emerging infectious diseases becoming an increasingly salient element of scientific collaboration and diplomatic activity, and involving participation by a wide variety of actors. These include philanthropists, activists, humanitarians, medics, corporate leaders, lawyers, technology entrepreneurs, religious groups, social movements and policy gurus, meaning that, while states remain central actors, global health governance is by no means an entirely state-centric business.

These changes have also been recognised and to some extent embraced by policy makers in their own reflections on, and interventions in, the art of global health governance. In 2007 Margaret Chan, director-general of the World Health Organization (WHO), stated when introducing a report aiming to define governance approaches in this new context, that 'The new watchwords are diplomacy, cooperation, transparency and preparedness' (World Health Organization, 2007, p. 2). While also stressing the importance of solidarity, Chan thus sought to define international public health security, or global health security as it has often been termed, as a rational, open, problem-solving activity, in contrast with the usual kinds of power struggles defining international affairs.

The emergence of infectious diseases as a major category of global health problems has been disruptive of existing patterns of governance. The international legal theorist David Fidler (2003) has argued that the governance of infectious diseases prior to the 1990s can be characterised as 'Westphalian', centred around the logic of sovereign states more or less cooperating internationally on a restricted set of global health problems. Certain international health initiatives notwithstanding (the elimination of smallpox declared by WHO in 1979 being the outstanding example), international initiatives on infectious disease tended to take place within the context of bilateral aid programmes, under the rubric of international development. While this account to some extent underplays the earlier development of 'global' health networks and practices (Bashford, 2006a; 2006b), it is true to say that in the decades prior to the 1980s, matters concerning the surveillance and control of infectious diseases did not form a recurring topic on the international political agenda. However, in the period since the 1980s, infectious diseases have been more disruptive of international relations, and the precise way in which multiple actors should cooperate in order to address global health problems has at times been intensely contested, with consequences for the effectiveness of disease control.

Furthermore, the terms in which 'global' is posed as a problem have themselves also been contested. Crucial here are the positions that different actors and analysts take with regard to the international political economy of health and especially the implications of *neoliberal* globalisation for global health. As Matthew Sparke (2009) has argued, from the 1980s onwards the dominant narratives of global health, that is, those emerging from the economic, political and intellectual centres in the global North, came to emphasise the role of markets, liberalisation, privatisation, cost-cutting and cost-recovery as the preferred route for health sector reform. In the neoliberal prescription, marketisation, welfare reform and structural adjustment would set the conditions for transitions to wealthier and healthier societies world-wide. It is disseminated via the World Bank (which in the 1980s overtook the WHO as the dominant force in international health policy formation), the World Trade Organisation (WTO) and associated think tanks and policy intellectuals. This prescription proved deeply problematic for many global South countries, whose health and welfare systems were undermined, opening space for newly emerging diseases, notably HIV and AIDS, and re-emerging epidemics, notably cholera and tuberculosis, to gain a stronger hold. As a result, health activists increasingly teamed up with campaigners against corporate-driven globalisation agendas to demand an end to neoliberalisation and the adoption of alternative policy agendas. As struggles over globalisation grew, and as evidence of the deleterious effects of neoliberalism on health in many global South countries was marshalled, the market fundamentalist approach was in many respects adjusted to allow for investment in tackling diseases of poverty that were understood to be holding affected populations back from joining the global economy, an approach that Sparke describes as 'market foster care'. But, as the remainder of the chapter will show, struggles over the political-economic terms of global health governance, struggles that are embedded within and conditioned by long term disparities in state and corporate power and wealth, have continued to surface.

All of this means that the shift towards a world whose health is managed by constellations of experts using new technologies and acting according to scientific and economic rationality remains far from being achieved, nor is it realistic to imagine that it can ever be. As will become clear in the following discussion, and despite aspirations for the technocratic administration of global health, it is for a variety of reasons simply not possible to eliminate politics, and more specifically geopolitics, from global health governance. More important perhaps are the ways in which (geo)political interests, conflicts and failures can feed back into the constitution of global health itself, influencing which global health problems are given greatest priority and whose needs are met to the greatest extent. It is thus important to account for the ways in which global health is geopolitical.

Although dozens of emerging diseases have been identified since the 1980s, the emerging diseases worldview (as it was termed by historian of medicine Nick King (2002)) has created a picture of a generalised threat from emerging pathogens. The geopolitics of global health have been animated by a series of epidemics that have resembled *crises of circulation*, as conceptualised by Foucault, in which there is the threat of a sudden escalation of transmission and spread beyond the capability of control measures. Such crises occasion intense phases of collaboration, innovation, contestation and change in apparatuses of security, which then serve as frames for further rounds of global health politics and the context for responses to the next crisis. In the remainder of the chapter, I will briefly trace the course of four epidemics that have played an important part in driving the transformation of international health into global health governance, the geopolitical tensions that have arisen out of them and how these have been addressed.

The humanitarian security emergency of HIV and AIDS

It is likely that the human immunodeficiency virus (there are in fact several distinct strains of HIV) crossed from primate populations into human ones in West Africa several times over the course of the last hundred years or so. Intriguingly, the earliest evidence of HIV occurring in humans comes from the time when Belgian colonialism was intensifying human–animal–environment transactions and circulations in the Congo in the late nineteenth and early twentieth centuries (Timberg and Halperin, 2012). But it is with the intensified patterns of transaction and circulation that are often described as contemporary globalisation, from the 1970s onwards, that HIV emerged definitively into human populations, leading eventually to what by the late 1990s was being described as the AIDS *pandemic* (a generalised outbreak affecting all areas, in this case of the globe, compared with a more localised epidemic). HIV is the virus that causes acquired immune deficiency syndrome (AIDS), resulting in the body's defences collapsing under an onslaught of opportunistic infections, usually some 5–10 years following infection). HIV was subsequently spread into many populations worldwide via sexual contact, unsafe blood-handling practices and injecting drug use. In many countries HIV and AIDS described concentrated epidemics, generally affecting the most at-risk populations. By contrast, across many regions of sub-Saharan Africa,

HIV proliferated into generalised epidemics, hitting people aged 15–40 particularly hard and spreading via 'vertical transmission' from mothers to children.

In the global North, activist and medical mobilisation led to a rapid political response in the face of widespread prejudice against the groups most affected by HIV and AIDS and fear at the potential extent of a new, and initially poorly understood, infectious disease. While effective medical countermeasures in the form of treatments or vaccines were not forthcoming, information about how HIV spread generally led to major preventive public health campaigns (which took different forms in different countries) and the provision of care for those dying of AIDS. However, due to a lack of medical and health infrastructures, HIV spread mostly unchecked among the most at-risk populations in sub-Saharan Africa with few exceptions. By the mid-1980s, a small number of global health activist experts, centred around the late Jonathan Mann, were pushing for a greatly expanded international response to HIV and AIDS, but were thwarted by bureaucratic politics and political inertia. At a time of retrenchment and structural adjustment, few power holders in international health welcomed the idea of a new infectious disease demanding new funding and programmes. Campaigners lacked a sufficiently convincing epidemiological picture of epidemics worldwide. Therefore epidemics progressed and multiplied through the 1990s, taking hold among new populations, especially in South America, Southeast Asia and the former Soviet Union, and especially where structural adjustment had undermined health care systems, increased poverty and amplified social dislocation.

By comparison with viral diseases like influenza and severe acute respiratory syndrome (SARS), HIV is slow. It may be many years after infection that viral loads increase, bringing the onset of AIDS and, without effective intervention, death. But by the mid-1990s, it was becoming apparent that HIV and AIDS were responsible for rapidly growing morbidity and mortality in the most affected populations of sub-Saharan Africa and were becoming increasingly pronounced in many other places. In 1996, a scientific and medical breakthrough was made with the discovery that several anti-HIV drugs in combination could achieve dramatically greater results than if used singly. The discovery of Highly Active Antiretroviral Therapy (HAART, often shortened to ART) was an important break point in the response to HIV/AIDS, but new innovative formulations were held under patent by global North pharmaceutical companies. While most global North countries could afford to introduce government-funded treatment programmes, thereby potentially extending the life of people living with HIV for many years, epidemics in poor countries went untreated.

It was at this point, around 2000, that the global politics of HIV and AIDS shifted from inaction to contestation. With epidemiological projections emphasising the growing scale and scope of HIV and AIDS epidemics in the most dramatic terms, and with the disease increasingly described as a global humanitarian emergency with security implications (it was speculated in a series of think tank reports that uncontrolled HIV/AIDS epidemics might destabilise whole regions or strategically significant states), political alignments began to shift.

AIDS activists called for pharmaceutical companies to relinquish their patents and for generic ART drugs produced by Indian companies to be made available internationally (Smith and Siplon, 2006). A number of countries (Brazil, South Africa, Thailand) made moves to initiate government-funded treatment programmes predicated upon generic or patent-free drug prices, and global political figures such as the UN secretary-general began to call for rich countries to set up a fund to support treatment outside the global North. As drug companies and the global North countries backing them relented, conceding the patents issue for first-line ART, growing funds were allocated to new international vehicles (the most important being the Global Fund to Fight HIV/AIDS, Tuberculosis and Malaria) and bilateral programmes (the largest being the US President's Emergency Plan for AIDS Relief). With this influx of new pharmaceutical treatments, political will and financial backing, infrastructures to deliver ART to in-need populations were, by the standards of international health, rapidly scaled up and it was claimed that by 2011 more than half of people in need of access to ART in developing countries were receiving it and access to services for the prevention of transmission and care of people living with HIV had also been increased (UNAIDS, 2012).

From the perspective of biosecurity it is also necessary to note that there have also been tensions surrounding the freedom of people living with HIV to travel internationally. Following the emergence of HIV and AIDS as a global disease, many countries introduced travel restrictions on people living with HIV. While UNAIDS, the agency established in 1996 to coordinate the response across UN institutions, has asserted that there are no public health grounds for restricting the movement of people living with the disease, and many countries have eliminated bans, continuing stigma surrounding HIV and AIDS and the continuing geopolitical appeal of control measures centred on the borders of a state imagined as sovereign have meant that, despite progress, travel restrictions remain in place in some countries (International Task Team, 2009).

The tensions surrounding international responses to HIV and AIDS are worth recounting at some length because they predate and in some respects prefigure later tensions around SARS and influenza in relation to intersections between techniques of government and questions of international political economy. Crucially, while the burden of infectious diseases remains overwhelmingly concentrated in global South countries, the research, development and production capabilities necessary to produce treatments and vaccines (it is important to note that no vaccine for HIV has yet been found) for new infectious diseases have been concentrated in global North countries, whose governments have since the early 1990s been promoting stringent new intellectual property regimes that shore up their position in the international political economy and prevent drugs from becoming available internationally until patents are relinquished and prices fall under generic competition.

But there are important differences too. One key difference between HIV and diseases like SARS and influenza has to do with temporality. While there is a years-long lag between HIV and the onset of AIDS, the gap between infection and symptoms in the case of other viral diseases can be measured in hours or days. Furthermore, the 'long-wave' temporalities of HIV/AIDS (Whiteside, 2008) have conditioned patterns of scientific research, industrial production, social mobilisation, political contestation and policy formation. The lag between infection and symptoms has meant that a proportion of people living with HIV have been able to become activists and that even a belated international response has been able to save many lives through a years-long (and still incomplete) process of scaling up access to treatment and other services. Things are different with influenza and influenza-like diseases, which spread through ordinary social contact and incapacitate people and organisations over a time span of days. While HIV/AIDS was a slow-burn event that was only dramatised as an emergency in international politics after more than a decade of activism and research, the effects of influenza and influenza-like microbes can become apparent at population level across countries and continents within days and weeks.

Influenza, SARS and the geopolitics of global health security

Virologists, microbiologists and other scientists in the US began calling for more concerted public policy on emerging infections in the late 1980s, with a series of conferences and reports emphasising the threat they posed to the US population, economy and, increasingly, national security (Lederberg et al., 1992; National Intelligence Council, 2000; King, 2002). In 1995, the World Health Assembly (WHA), the governing body of the WHO, adopted a resolution calling for the revision of the International Health Regulations (IHRs), the main instrument governing how states should respond to outbreaks of infectious diseases. Originating in the 1850s, the IHRs required international notification of only a small number of specified diseases, and it was increasingly apparent that this was an inadequate basis for global health governance. In 1996, President Clinton issued a Presidential Decision Directive (Office of Science and Technology Policy, 1996) setting out a series of policy goals to expand the USA's role in global health and to improve national and international surveillance and response capabilities. Moves to give greater emphasis to emerging infections were fuelled by increased Western media coverage of epidemics of exotic-sounding diseases like hemorrhagic fever, Lassa fever, Marburg virus, Ebola, West Nile virus and plague. In two highly influential books, the journalist Laurie Garrett (1995; 2000) dramatised the threat of emerging infections and lambasted political authorities for their failure to act.

An important set of events took place in Hong Kong in 1997, when an epidemic of H5N1 avian influenza in poultry crossed over into the human population, causing 18 cases of flu, of which six were fatal (Sims et al., 2003). This was ominous for a number of reasons. First, it demonstrated the potential for avian flu viruses to cross into human populations in a context of intense human–animal interactions and transactions and one that, with demand for meat increasing rapidly with economic growth, looked set to become more significant. Second, the apparent fatality rate from these infections was one in three. At this stage, no human-to-human transmission of the flu virus – a condition for a human pandemic – was detected. But should the genetic composition of the virus be reassembled in the process of transmission to humans in such a way as to facilitate easy onward transmission, the consequences of such a high fatality rate could be devastating. The severity of the threat posed by H5N1 in the 1997 Hong Kong outbreak was reflected in the control measures instituted in an attempt to end it: total eradication of the poultry population, or more than 1.5 million chickens.

Following the Hong Kong outbreak, efforts increased to upgrade international systems for the monitoring of animal and human populations and the detection of unusual outbreak-type events. In 2001, the WHA adopted a resolution on global health security that, in recognition that international transparency and collaboration on disease outbreaks had often been patchy, called upon member states 'to participate actively in the verification and validation of surveillance data and information concerning health emergencies of international concern' (World Health Assembly, 2001). In 1997 WHO had started gathering reports of unusual public health events informally from non-governmental sources, allowing it to some extent to circumvent conventional reporting channels, which were often slow and subject to political vicissitudes (reporting an epidemic could adversely affect trade and tourism), potentially allowing WHO greater leverage over member states in a crisis. Notable among these new sources of data was the Global Public Health Intelligence Network (GPHIN), a Canadian government initiative that, instead of looking for confirmed epidemiological reports, used webcrawlers to search online news for reports of unusual public health events.

From its partnership with GPHIN, WHO learned of reports of a strange, pneumonia-like disease circulating in southern China in late 2002 and was able, in the face of assertions that the situation was under control, to pressure the Chinese government to investigate the outbreak and to share its data. By early 2003, however, SARS, a new respiratory disease caused by a coronavirus that had originated in a population of civet cats, had spread to a number of other countries, affecting hospital workers especially severely. Amid growing tensions with China, WHO angered the Canadian government by unilaterally issuing a travel advisory warning people not to visit Toronto, which was experiencing its own SARS epidemic. Toronto's economy, heavily reliant on international convention visitors and tourism, suffered a devastating blow and the Canadian government sought to restore public confidence. Coinciding with the run-up to the US and UK invasion of Iraq, the anxieties surrounding the SARS epidemic were amplified by news networks taken with narratives of global crisis. Though in global public health terms SARS was a small event, with 8,096 probable cases and 774 deaths attributed by WHO to the disease, SARS was widely considered to be a 'wake-up call' and further catalysed efforts to reform global health governance (Fidler, 2003).

In 2005 the WHO agreed revisions to the International Health Regulations, which now, as well as a specified list of diseases that member states were expected always to report to WHO, covered all 'public health emergencies of international concern' (World Health Organization, 2005). States were expected to designate an IHR liaison and to upgrade their disease surveillance capabilities to minimum accepted levels. However, the IHR process included no discussion of financial assistance or solidarity clauses for states that would struggle to implement these measures on their own.

The move towards a global health security paradigm for outbreak events has involved significant shifts in the rationalities and technological organisation of global health governance. First, the WHO moved towards a use of the term 'security' – hitherto associated with militarised national security institutions and paradigms of the sort that WHO had striven to avoid – in order to maintain its political neutrality as a UN agency. The embrace of 'security' both dramatised global health politically and signalled that new kinds of rationalities and technologies would henceforth be in play. Second, there has been a shift from 'public health information' to 'public health intelligence'. Besides being another indication of the seepage of national security terminology into the public health field, this signals a shift from a reliance on confirmed reports of disease events to an inferential approach based on the analysis of diverse types of information and data that may or may not suggest the presence of a disease outbreak. Thus global health security shifts onto a virtual level, where the potential existence of a phenomenon can become grounds for action (Braun, 2007). Third, there is also a shift in the spatio-temporalities of global health governance, as by utilising media and, increasingly, social networking technologies, public health agencies seek to get as close to epidemiological time-spaces as possible in order to interrupt them. In these ways, public health has come to resemble and in some cases intertwine with the work of intelligence and security agencies with which they increasingly share rationalities and technologies. According to Lorna Weir and Eric Mykhalovskiy (2010), WHO was forced to reassess its relationship with GPHIN after it began supplying reports of unusual events to intelligence and military organisations, and, according to David Fidler (2005), tensions arose during the IHR negotiation process when one state (unnamed, but which can be inferred as the USA) pushed for the new regulations to include provisions for any samples associated with unusual events (including chemical, biological and radiological incidents) to be transferred to international jurisdiction, a move that was interpreted as an attempt to make WHO and public health an instrument of military biosecurity.

These tensions broke into the open during the course of 2006 and 2007 when, with increasing cases of H5N1 avian influenza, including animal-to-human transmission, and reports of human-to-human transmission in a number of Southeast Asian countries, Indonesia announced that it was ceasing its sharing of flu samples with the WHO. Since 1948, countries had freely shared samples of influenza with WHO collaborating centres, which would synthesise a vaccine that could then be produced industrially for distribution in time for the seasonal onset of flu. But, amid reports that WHO centres had, in contravention of the protocol governing sharing, given samples to private companies, who were pursuing patents on flu-derived materials, Indonesia withdrew from the system. The subsequent dispute highlighted the contentious and interconnected nature of the techniques of security and international political economies at work in global health.

In apparently ill-judged comments, and referring to a US military biomedical research facility in Jakarta, Indonesia's health minister claimed that the flu samples were being used by the USA to develop biological weapons. Though flatly denied, the comments played to much broader anxieties about US foreign policy and its interventions in Muslim countries as part of the 'war on terror'. Amid the ensuing diplomatic storm, Laurie Garrett and Richard Holbrooke (Holbrooke and Garrett 2008), another influential figure in global health politics, wrote an op-ed in the *Washington Post* condemning Indonesia's stance, urging Washington to exert diplomatic pressure and demanding immediate compliance and transparency in the name of global health. With EU countries and others in the global North invoking global health security (a concept used by the WHA in its 2001 resolution), Indonesia, Brazil, India and other middle-income countries began to question the meaning of the term itself, arguing that, while the WHO had been using the term, its content and use had never been agreed at the WHA (Tayob 2008).

There was much more to the dispute than the form of transparency apparently required in the name of global health security. Indonesia and its allies claimed that behind the virus-sharing dispute lay profound issues of global health equity. While being urged to share viral samples, Indonesia and other countries, it was claimed, would not be able to access any medications (vaccines or anti-viral agents) that might be derived from them, while private corporations based elsewhere would benefit from revenues derived from patented drugs. The claim regarding profits was only partially valid. Though demand for on-patent anti-virals can make them financially attractive (the share price of companies holding rights in Tamiflu, for which some efficacy against avian flu was claimed, increased sharply), the economic model under which influenza vaccines are produced is not particularly profitable. But the point on access to medicines was well founded: global vaccine and anti-viral production capacity was far short of being able to secure global population coverage during a flu pandemic, and global North countries hosting vaccine and drug producers were most able to secure and pay for access. As concern about the potential destructiveness of an influenza pandemic was peaking, Indonesia threw the whole basis of the global health security system into question.

The dispute turned towards law. With global North countries invoking global health security, Indonesia cited the Convention on Biological

Diversity, which grants countries sovereignty over biological materials originating on their territory. While legal experts argued that the IHRs took priority, political agreement on a new framework for virus sharing was not forthcoming. In Spring 2009, however, tensions over virus sharing were overridden by the outbreak of a new epidemic, this time of H1N1 influenza originating in swine populations traced to Mexico. With early reports suggesting a high case-fatality rate and outbreaks rapidly spreading far beyond Mexico, WHO invoked the IHRs and declared an early stage pandemic, initiating a global public health response, with many countries introducing a range of control measures. With anxieties growing that H1N1 might be the long-feared pandemic of highly transmissible, highly lethal influenza, governments started to scramble to secure access to vaccine and anti-viral drug supply lines, and all production capacity was rapidly obtained by the global North countries hosting pharmaceutical and vaccine-producing companies. Though tensions over the sharing of samples of H1N1 did not arise, disquiet rapidly grew about the equity of arrangements for access to vaccines and anti-virals. It was only once it had become clear that a single dose of vaccine would suffice to immunise adults and that H1N1 would be much milder than had been feared in terms of its fatality rate that global North countries agreed to donate vaccines to international stockpiles accessible by other countries (Fidler, 2010).

At the diplomatic level, agreement was reached in spring 2011 on a new framework for sharing influenza viruses, which requires WHO member states, national influenza laboratories and private corporations to work together to increase access to vaccines, anti-virals and diagnostic kits for low-income countries and contains provisions for donor countries and corporations to contribute financially to this effort (Schnirring, 2011; WHO, 2011). Efforts are also being made to increase global vaccine production capability. However, there continues to be widespread scepticism over whether the agreement on virus sharing will hold in the event of another pandemic in the face of states' actors seeking to assert their sovereignty and protect their populations by whatever means possible.

Conclusions

Though developed primarily in relation to the arts of government within a state, Foucault's discussions of governmentality provide a useful set of ideas for considering the ways in which rationalities and technologies of security have been developed in the field of global health over the course of the three decades since the identification of HIV as the virus that causes AIDS. But while his work provides a useful conceptual vocabulary to describe the formation, operation and dynamics of apparatuses of security, they are less well attuned to the struggles over equity to which they often give rise. Though they are attuned to the ways in which conditions of liberal and neoliberal freedom are produced by political technologies, they are less well placed to identify the ways in which those political technologies are conditioned by disparities in property and power. Hence I have focused on the tensions in global health governance that have emerged at the intersections between technologies of security and the formation of the international political economy of health.

While global health governance is inevitably guided by scientific analysis and economic calculations, the role of power and politics in shaping priorities and policy programmes is frequently disavowed. However, the cause of more equitable and effective global health governance requires that the political tensions under-lying global health, whose lack of solution compromises the prospects of achieving health for all, are brought forward and made the subject of analysis and delibera-tion. While globalisation undoubtedly increases common exposure to microbial threats and therefore necessitates collective action, the global health field is fraught with tensions and dilemmas. As David Fidler (2010) has observed, invok-ing moral arguments about global health inequity is unlikely to effect the kinds of change that are widely considered to be desirable. The example of HIV/AIDS suggests that when social movements, scientific expertise and political-economic forces align in particular ways around an epidemic, changes that were previously hard to envisage are possible. What is unknown is whether the next pandemic will afford such time and space for action.

References

Bashford, A., 2006a. Global biopolitics and the history of world health. *History of the Human Sciences*, 19(1), pp. 67–88.

Bashford, A. (ed.), 2006b. *Medicine at the Border: Disease, Globalization and Security, 1850 to the Present*, Basingstoke, UK and New York: Palgrave Macmillan.

Braun, B., 2007. Biopolitics and the molecularization of life. *Cultural Geographies*, 14(1), pp. 6–28.

Elbe, S., 2009. *Virus Alert: Security, Governmentality, and the AIDS Pandemic*, New York: Columbia University Press.

Fidler, D. P., 2010. Negotiating equitable access to influenza vaccines: global health diplomacy and the controversies surrounding avian influenza H5N1 and pandemic influenza H1N1. *PLoS Med*, 7(5), p. e1000247.

—— 2005. From international sanitary conventions to global health security: the new international health regulations. *Chinese Journal of International Law*, 4(2), pp. 325–392.

—— 2003. SARS: political pathology of the first post-Westphalian pathogen. *The Journal of Law, Medicine and Ethics*, 31(4), pp. 485–505.

Foucault, M., 2007. *Security, Territory, Population: Lectures at the Collège de France 1977–1978*, Houndmills: Palgrave Macmillan.

—— 1980. The confession of the flesh. In *Power/Knowledge: Selected Interviews and Other Writings*, Hemel Hempstead, UK: Harvester Wheatsheaf, pp. 194–228. Available at: http://foucaultblog.wordpress.com/2007/04/01/what-is-the-dispositif (accessed 3 September 2012).

Garrett, L., 2000. *Betrayal of Trust: The Collapse of Global Public Health*, New York: Hyperion.

—— 1995. *The Coming Plague: Newly Emerging Diseases in a World Out of Balance*, London: Penguin.

Holbrooke, R. and Garrett, L., 2008. 'Sovereignty' that risks global health. *Washington Post*. Available at: www.washingtonpost.com/wp-dyn/content/article/2008/08/08/AR200808 0802919.html (accessed 3 September 2012).

International Task Team, 2009. *Mapping Restrictions on the Entry, Stay and Residence of People Living with HIV*, Geneva: UNAIDS. Available at: www.unaids.org/en/media/ unaids/contentassets/dataimport/pub/report/2009/jc1727_mapping_en.pdf (accessed 20 December 2012).

King, N. B., 2002. Security, disease, commerce ideologies of postcolonial global health. *Social Studies of Science*, 32(5–6), pp. 763–789.

Lakoff, A. and Collier, S. J., 2008. *Biosecurity Interventions: Global Health and Security in Question*, New York: Columbia University Press.

Lederberg, J., Shope, R. and Oaks, S. (eds), 1992. *Emerging Infections:Microbial Threats to Health in the United States*, Washington DC: National Academies Press.

National Intelligence Council, 2000. *The Global Infectious Disease Threat and Its Implications for the United States*, Washington DC: NIC. Available at: http://cdm16064. contentdm.oclc.org/utils/getfile/collection/p266901coll4/id/1805/filename/1695.pdf (accessed 4 September 2012).

Office of Science and Technology Policy, 1996. *PDD NTSC-7 Addressing the Threat of Emerging Infectious Diseases*, Washington, DC: The White House. Available at: www.fas. org/irp/offdocs/pdd_ntsc7.htm (accessed 21 December 2012).

Schnirring, L., 2011. WHO group finalizes landmark pandemic virus-sharing agreement. *CIDRAP*. Available at: www.cidrap.umn.edu/cidrap/content/influenza/panflu/news/ apr1811sharing-br.html (accessed 3 September 2012).

Sims, L. D., Ellis, T. M., Liu, K. K., Dyrting, K., Wong, H. et al., 2003. Avian influenza in Hong Kong 1997–2002. *Avian diseases*, 47(3 suppl.), pp. 832–838.

Smith, R. A. and Siplon, P. D., 2006. *Drugs Into Bodies: Global AIDS Treatment Activism*, London: Praeger.

Sparke, M., 2009. Unpacking economism and remapping the terrain of global health. In *Global Health Governance: Crisis, Institutions and Political Economy*, Houndmills, UK: Palgrave Macmillan, pp. 131–159.

Tayob, R., 2008. WHO board debates 'global health security', climate, IPRs. *Third World Network*. Available at: www.twnside.org.sg/title2/intellectual_property/info.service/ 2008/twn.ipr.info.080101.htm (accessed 21 December 2012).

Timberg, C. and Halperin, D., 2012. *Tinderbox: How the West Sparked the AIDS Epidemic and How the World Can Finally Overcome It*, New York: Penguin Press.

UNAIDS, 2012. *Global Report*, Geneva: UNAIDS. Available at: www.unaids.org/en/ media/unaids/contentassets/documents/epidemiology/2012/gr2012/20121120_UNAIDS_ Global_Report_2012_en.pdf (accessed 20 December 2012).

Weir, L. and Mykhalovskiy, E., 2010. *Global Public Health Vigilance: Creating a World on Alert*, New York: Routledge.

Whiteside, A., 2008. *HIV: A Very Short Introduction*, Oxford: Oxford University Press.

WHO, 2011. Landmark agreement improves global preparedness for influenza pandemics. *World Health Organization*. Available at: www.who.int/mediacentre/news/releases/2011/ pandemic_influenza_prep_20110417/en/index.html (accessed 21 December 2012).

World Health Assembly, 2001. *Global Health Security: Epidemic Alert and Response*, Geneva: World Health Organization. Available at: http://apps.who.int/gb/archive/pdf_files/WHA54/ea54r14.pdf (accessed 20 December 2012).

World Health Organization, 2007. *A Safer World: Global Public Health Security in the 21st Century*, Geneva: WHO. Available at: www.who.int/entity/whr/2007/whr07_en.pdf (accessed 20 December 2012).

—— 2005. *International Health Regulations (2005)*, Geneva: WHO.

10

BIOSECURITY AND BIOTERROR

Reflections on a decade

Brian Rappert and Filippa Lentzos

Introduction

This chapter focuses on the interpretation of the 'biosecurity' threat as bioterrorism. It examines how the political focus on the threat from disease since the early 1990s – but particularly post-September 11th and the ensuing anthrax attacks – was initially on high-consequence 'superterrorism' events, but in more recent years has broadened to include threats from sabotage, negligence, accidents, unintended consequences and natural disease outbreaks. The chapter goes on to illustrate how the evolving political understanding of bioterrorism is complemented by tensions and contradictions in the responses and initiatives undertaken to address the threat.

The unprecedented terrorist attacks on New York and Washington on September 11th 2001 led the US Congress to establish the 9/11 Commission to investigate the circumstances surrounding the attacks. This commission determined that 'the greatest danger of another catastrophic attack in the United States will materialize if the world's most dangerous terrorists acquire the world's most dangerous weapons' (9/11 Commission, 2004, p. 380). It recommended the creation of a follow-on commission to further examine the threat posed by this nexus of international terrorism and the proliferation of chemical, biological, radiological and nuclear (CBRN) weapons (also often termed 'unconventional' weapons or 'weapons of mass destruction' (WMD)). The Commission on the Prevention of WMD Proliferation and Terrorism as it was termed, or 'WMD Commission' for short, declared in its 2008 *World at Risk* report that, despite prevention efforts to date, the margin of safety against WMD terrorism was shrinking, not growing. It noted that 'unless the world community acts decisively and with great urgency, it is more likely than not that a weapon of mass destruction will be used in a terrorist attack somewhere in the world by the end of 2013' (WMD Commission, 2008, p. xv). While its mandate was to examine the full range of challenges posed by terrorist use of CBRN weapons, the WMD Commission concluded early in its deliberations that it would focus solely on the two weapons that have the greatest potential to kill the most massive numbers: biological and nuclear weapons. Of these two, the WMD Commission concluded that, 'terrorists are more likely to be

151

able to obtain and use a biological weapon than a nuclear weapon' (WMD Commission, 2008, p. xv). It declared that a biological attack that inflicts mass casualties is more likely in the near term because of the greater availability of the relevant so-called 'dual-use' materials, equipment and know-how, which are spreading rapidly throughout the world.

Ten years on from September 11th the WMD threat has yet to manifest itself in a manner many politicians, lobbyists or journalists thought it would: through mass disruption to vital financial, communications, and transportation systems and thousands of deaths. Indeed, there has been no mass destruction from terrorist use of CBRN weapons either before or after September 11th. In terms of biological weapons, there have only been three confirmed attempts to use them against humans for terrorist purposes in recent decades: the 1984 use of salmonella by the Rajneesh cult in Oregon, the 1990–1995 attempted use of botulinum toxin and anthrax by the Aum Shinrikyo cult in Tokyo and the 2001 'Amerithrax' distribution of a high-quality dry-powder preparation of anthrax spores attributed to the biological weapons scientist Bruce Ivins. In addition to these incidents, there have been thousands of hoaxes or false claims that a biological attack has been perpetrated. Some commentators argue that hoaxes, which are ideologically motivated and intended to cause fear and intimidation, should be counted as bioterrorism incidents. Others argue they should not, as no actual biological agent is involved. Whether or not we count these as bioterrorism incidents, what is clear is that the debate about the threat of bioterrorism has had significant political salience.

On the 10-year anniversary of 9/11 and the Amerithrax attacks, the former US senators who chaired the WMD Commission, Bob Graham and Jim Talent, released a 'report card' on America's bio-response capabilities (WMD Center, 2011). The report card asserts that the threat is grave – it states, for instance, that 'Modern biotechnology provides small groups the capabilities for a game-changing bio-attack previously reserved to nation-states. Even more troubling, rapid advances in biotechnology, such as synthetic biology, will allow small teams of individuals to produce increasingly powerful bioweapons in the future' (WMD Center, 2011, p. 14). This sort of rhetoric is to be expected, since both former senators tend to rather extreme views on the threat we face from bioterrorism. But what is new in their discourse is the assertion that we must also prepare ourselves against the threat of naturally occurring diseases: 'Today we face the very real possibility that outbreaks of disease, naturally occurring or man-made, can change the very nature of America – our economy, our government and our social structure' (WMD Center, 2011, p. 11).

In the rest of this chapter we consider how the security threat from disease and what are considered appropriate responses have changed over the last decade. In our concluding section we look forward and consider what the next 10 years might hold.

Ten years of threats

Biosecurity as bioterrorism

'Bioterrorism' is a relatively new concept that emerged during the early 1990s in the United States in reference to terrorists' potential access to biological weapons. Biological weapons themselves have a longer history, extending back to the interim between the two World Wars, when they started being developed and tested in state programmes (Wheelis et al., 2006; Guillemin, 2005a). Most of these programmes ceased when the Biological Weapons Convention (BWC) banning biological weapons came into force in 1972, with one notable exception: the Soviet Union. The Soviet Union did not believe in US president Nixon's claim to have renounced biological weapons, and while it publicly signed up to the Convention, it secretly began an extraordinary expansion of its biological weapons programme. The expansion coincided with scientific advances in genetically modified organisms and resulted in a new generation of very sophisticated weapons. During the Cold War, the threat from these weapons was held in check through a combination of the BWC and the threat of nuclear retaliation. In the last years of the Cold War, however, a new set of threats posed by rising Third World states and terrorists supported by these states began to be projected by some US security analysts, and among these threats were terrorists armed with biological weapons and other 'WMD'. As the Cold War faded, the threat of biological weapons from Third World states and terrorists hostile to the US began to replace the Soviet threat. Although little credible evidence existed at the time that such states or terrorists would or even could resort to biological weapons, the newly perceived threat became the driving force behind US preparedness and biodefense programmes of considerable institutional proportions (Guillemin, 2005b; Wright, 2007).

The 2001 anthrax letters, coming within weeks of the September 11th attack, changed the political significance of bioterrorism by an order of magnitude. Bioterrorism shot up the public 'risk portfolio' (Douglas and Wildavsky, 1982). President Bush made several statements stressing the severity of the threat. In one of his first statements on biological weapons he said: 'Since September 11, America and others have been confronted by the evils [biological weapons] can inflict. This threat is real and extremely dangerous. Rouge states and terrorists possess these weapons and are willing to use them' (The White House, 2001). The bioterrorists and their 'rogue-state patrons' were given a face when the US named its adversaries in a November 2001 statement to BWC members; they included Osama bin Laden and al Qaeda, Iraq, North Korea, Iran, Libya, Syria and Sudan.

Building a biodefence capacity rapidly became a critical national priority for the Bush administration. The Homeland Security Presidential Directive of April 2004 describes the administration's biodefence strategy. The document is classified, but the non-classified version, 'Biodefense for the 21st Century,' spells out its commitment to biodefence: 'The United States will continue to use all means necessary to prevent, protect against, and mitigate biological weapons attacks perpetrated against our homeland and our global interests' (NSPD-33, 2004). This has involved a vast array of biosecurity formations including threat- and vulnerability-assessment exercises, prevention-and-protection efforts, surveillance-and-detection programmes and response-and-recovery initiatives, as well as a significant expansion of the biodefence infrastructure and the training and funding of thousands of scientists, researchers and technicians to work on bioterrorism countermeasures (Lentzos and Rose, 2009; Lakoff and Collier, 2008; Lentzos, 2006).

The threat stemming from the nexus of international terrorism and the pro-liferation of WMD, accentuated by the 9/11 Commission, also became recognized outside the United States and rapidly materialized in the international risk portfolio and in security strategies. The international community's premier secur-ity forum, the Security Council of the United Nations, unequivocally condemned the September 11th attacks and affirmed that the attacks, and indeed any acts of international terrorism, constituted a threat to international peace and security. In early 2004, the Security Council unanimously adopted resolution 1540, in which it expressed that it is 'Gravely concerned by the threat of terrorism and the risk that non-State actors. ... may acquire, develop, traffic in or use nuclear, chemical and biological weapons', and obliged states to enact national legislation to ensure that terrorists do not obtain these weapons.[1]

The then secretary-general of the United Nations, Kofi Annan, used his address to BWC members at their 2006 quinquennial meeting to highlight bioterrorism and urged that the treaty should also deal with the threat from terrorists:

> In the [last] five years ... global circumstances have changed, and risks evolved. We see today a strong focus on preventing terrorism ... These changes mean that we can no longer view the Convention in isolation, as simply a treaty prohibiting states from obtaining biological weapons. ... we must also address terrorism and crime at the non-state and individual levels.[2]

By the time of the next quinquennial meeting of BWC members, in 2011, a large number of national statements made references to the bioterrorism threat, including Russia, China, India and the European Union, as well as Non-Aligned Movement states.[3] Not only had states included bioterrorism as part of their risk portfolios, obligations under Security Council resolution 1540 and the BWC ensured that they had also implemented measures to counter bioterrorism in national legislation. In its September 2011 report to the Security Council, the 1540 committee reported an upward trend by states in implementing legislation that 'has contributed to strengthened global non-proliferation and counter-terrorism regimes and has contributed to better preparing states to prevent proliferation of [WMD] to non-state actors'.[4]

One hundred and fifteen states had laws prohibiting the use of biological weapons; 112 states had laws prohibiting the manufacture, production and acquisition of biological weapons; 72 states had laws prohibiting the possession of biological weapons; 103 had laws prohibiting the stockpiling of biological weapons; 98 had laws prohibiting the development of biological weapons; 52 had laws prohibiting the transport of biological weapons; 104 had laws prohibiting the transfer of biological weapons; and 133 states had adopted enforcement measures related to the use, manufacture, acquisition, possession, stockpiling, development, transfer and transport of biological weapons.[5]

The political focus on the bioterrorism threat was sustained over the 2001–2011 decade by the perception that biological weapons are increasingly becoming accessible and affordable through scientific advances. Concern in the US has concentrated particularly on the nascent field of synthetic biology, where techniques to genetically modify biological agents could make it easier for terrorists to acquire dangerous viral pathogens, particularly those that are restricted to a few high-security labs (such as the smallpox virus), are difficult to isolate from nature (such as Ebola and Marburg viruses), or have become extinct (such as the Spanish influenza virus) (Lentzos and Silver, 2012).

Biosecurity as the threat from disease

In parallel to the focus on bioterrorism, a different conception of the threat of disease that links security with health, has also been gaining political traction. Over the last decade, the World Health Organization (WHO) became a new actor in the security world and exerted significant influence on how the threat from disease was perceived. A few months after September 11th and the Amerithrax attacks, the governing body of the WHO, the World Health Assembly, noted its serious concern 'about threats against civilian populations, including those caused by *natural* occurrence or *accidental* release of biological or chemical agents … as well as their *deliberate* use to cause illness and death' (emphasis added).[6] Its key message, reiterated over the years by the WHO through its publications and presentations to forums like meetings of the BWC, was that, whatever the cause of epidemics or emerging infectious diseases, the response to them will initially be the same. The threat from the deliberate use of biological weapons should therefore be thought of as part of a wider spectrum of threats that also include the threat of disease from natural outbreaks and accidental releases, and the key response to these threats should be to bolster the global public health system. Adding urgency to the WHO's call to strengthen public health capacity was a string of epidemics in the 2001–2011 period, including the Severe Acute Respiratory Syndrome (SARS) outbreak in 2003, the re-emergence of H5N1 avian influenza, and the novel H1N1 swine-origin influenza in 2009.

In the United States, where the concept of bioterrorism first emerged, the gradual grouping together of natural outbreaks, accidental releases and deliberate infections also manifested itself in government policies over the 2001–2011 decade. Indications of the link between security and health were evident already in the latter part of the Bush administration's strategies. For instance, in the 2006 National Security Strategy, deadly pandemics were included in the same category of threats to national security as terrorist acquisition of nuclear, chemical and biological weapons (The White House, 2006, p. 44). The National Strategy for Public Health and Medical Preparedness, first established in 2007 and built on principles from the administration's 2004 biodefence strategy, included terrorist attacks with weapons of mass destruction alongside naturally-occurring pandemics as examples of 'catastrophic health events', transforming the 'national approach to protecting the health of the American people against all disasters' (The White House, 2007, pp. 1–2).

The spectrum approach to disease became further ingrained in the Obama administration's security policies. President Barack Obama endorsed 'a comprehensive approach that recognizes the importance of reducing threats from outbreaks of infectious disease whether natural, accidental, or deliberate in nature', and he put forward a National Strategy for Countering Biological Threats that articulated his administration's vision for managing 'these evolving and complex risks' (National Security Council, 2009, foreword). The strategy prominently featured 'the full spectrum of biological threats', stressing that 'the rapid detection and containment of, and response to, serious infectious disease outbreaks – whether of natural, accidental or deliberate origin – advances both the health of populations and the security interests of states' (National Security Council, 2009, p. 4).

The head of the State Department's bureau of international security explained what this shift in political understanding of the bioterrorism threat has entailed operationally:

> Traditionally, our focus has been on hard security issues, and the use of a range of tools, from sanctions and export controls to UN Security Council Resolutions and negotiations, to prevent governments from acquiring weapons of mass destruction. In the last decade or so, that focus has changed, particularly with respect to the life sciences and biological weapons. … As dual-use capabilities spread and non-state actors seek ever-growing destructive potential, we've had to adopt a wider range of tools, and work much more closely with colleagues in the public health, law enforcement and life sciences communities. Raising awareness, building capacity, and influencing attitudes and intentions – both within governments and at the level of laboratories and individual scientists – is increasingly central to our work.[7]

A similar shift in emphasis on what the security threat from disease is and how best to respond to it can also be seen more internationally. At the close of the 2011 quinquennial meeting of BWC members, for instance, all states recognized 'that achieving the objectives of the Convention will be more effectively realized through greater public awareness of its contribution, and through collaboration with relevant regional and international organizations'.[8] It is to these collaborative efforts that we now turn.

Ten years of responses: the life sciences and biosecurity

Against the shifting threats identified in the previous section, here we turn to more specific national and international initiatives undertaken since 9/11. In particular, among the range of issues that could be discussed, we examine the unfolding policy developments associated with the intersection between research and biosecurity. As alluded to above, research has been identified as both a major source of concern in relation to the hostile use of pathogens as well as a central means of responding to threats. The dual quality attributed to science has meant governance approaches adopted in response to concerns about bioweapons, and bioterrorism especially, have been characterized by contradictions and tensions.

Funding and facilities

Perhaps nowhere is this more evident than in relation to biodefence funding in the US. Since 2001, government support for research and development work in this area has manifoldly multiplied. While, in 2001, the funding for civilian biodefence totalled $569 million, in 2012 it was projected to be in excess of $6.4 billion (Franco and Sell, 2011). Research has been a core component of this expansion, with funding in excess of $3 billion per year since 2004, much of it commissioned by the National Institutes of Health (Center for Arms Control and Non-proliferation, 2008). To undertake this work, the number of high containment labs has also dramatically increased. While, in 2004, there were 415 registered 'BSL-3' labs,[9] by 2010 there were nearly 1,500 (Kaiser, 2011).

The surge in funding and facilities has been justified on the basis of the importance of basic and applied research, as well as diagnostic and vaccine procurement, in staying ahead of biothreats. Two overall criticisms have been made against this rationale: (1) the programmes have not been effective at achieving their stated goals, and (2) they have been counterproductive. With regard to the first criticism of *ineffectiveness*, concerns have been voiced about the absence of resulting countermeasures and the lack of relevance of the commissioned research. With regard to the latter, much of the original funding was designated for the study of individual classic bioweapon agents (e.g. anthrax, smallpox, plague). This has been criticized in relation to the impracticality of responding to attacks by devising treatments for individual pathogens, as well as the need for biodefence funding to serve public health goals of the kind aligned with the broader conceptions of biosecurity raised in the previous section (Hayden, 2011). One response since 2007 has been a greater attention to 'broad-spectrum' priorities that could be useful against multiple pathogens as well as in defensive and non-defensive health applications. In relation to facilities, the coherence of the building of high containment labs has been questioned. The United States Government Accountability Office (2009) stated it was not possible to find:

> a government wide strategic evaluation of future capacity requirements set in light of existing capacity; the numbers, location, and mission of the laboratories needed to effectively counter biothreats; and national public health goals [...] Furthermore, since no single agency is in charge of the expansion, no one is determining the aggregate risks associated with this expansion.

With regard to the second charge of being *counterproductive*, it has been argued not only that biodefence has been over-prioritized in the US, but that the work undertaken is muddling the boundaries of what counts as permissible for national defence, that it raises the possibility of accidental releases, that it has led to the proliferation of dangerous knowledge and that it increases the potential for intentional theft and misuse of materials (see e.g. Sunshine Project, 2004; Enserink and Kaiser, 2005; Leitenberg et al., 2004; Klotz and Sylvester, 2009). Such evaluations were given additional weight in 2008 when the Federal Bureau of Investigation announced that the suspected perpetrator of the anthrax attacks worked within the US Army Medical Research Institute of Infectious Diseases.

While the US is unrivalled in relation to the magnitude of its expansion in biodefence-related funding, internationally the number of new high-level biocontainment laboratories is also increasing rapidly. In large part this is because of funding by donors such as the World Bank, the Asian Development Bank and the Global Partnership, as well as funding streams made available as part of national agencies in Japan, the US, Canada and elsewhere (National Research Council, 2012). The stated rationale for such building work has followed the shifting appraisals made of the threats from state programmes, bioterrorism, accidental release and infectious diseases over the last decade.

Laboratory governance

As part of attempts to reduce bioattacks, including those from 'insiders', since 2001 countries such as the US, Australia, France, Japan and the UK have introduced new legislation specifically related to laboratories working with agents that are deemed to pose security risks. For instance, the 2001 USA Patriot Act outlawed so-called 'restricted persons' from possessing designated 'select agents'. Those persons included foreign nationals from countries said to be supporting terrorism (at the time of writing that list consisted of Cuba, Iran, Sudan and Syria), convicted felons and others. The 2002 US Public Health Security and Bioterrorism Preparedness and Response Act, known as the Bioterrorism Act, required labs to devise plans to avoid the accidental or deliberate release of sensitive materials as well as background checks for those working with agents of concern. Enhanced controls were also introduced for domestic and international transfers of such agents. Under the 2009 Executive Order 13486, titled *Strengthening Laboratory Biosecurity*, the US government is currently considering additional measures.

Not all countries have opted for the introduction of new rules and regulations on the access to pathogens in relation to national concerns and international obligations. Some criticized initial conceptualizations of biosecurity offered within the international stage in the years after 2001 for focusing too much on laboratory governance vis-à-vis bioterrorism at the expense of treating disease as a security threat (Revill and Dando, 2007). As noted in the introductory chapter to Rappert and Gould (2007), many other such differences in policies and practice that index contrasting assessments of threats can be identified (see as well Barr, 2007; Tucker 2007).

Experiment and manuscript reviews

One type of security response that has animated much debate has been the review of civil research proposals and publications for their potential malign use. A number of high-profile scientific experiments[10] have been flagged as demonstrating the potential that life sciences have to be misused and have also been important in maintaining attention on biosecurity as bioterrorism. Since 2003 a number of funders, publishers and organizations based in North America and Europe have introduced oversight processes to assess the risks and benefits of individual proposals or manuscripts to determine whether they need to be modified or not taken forward (Rappert, 2008). As an example, in 2003, a group of leading science journals agreed general guidelines for modifying and perhaps rejecting manuscripts where 'the potential harm of publication outweighs the potential societal benefits' (Journal Editors and Authors Group, 2003, p. 1464). As another example, in 2006 a National Science Advisory Board on Biosecurity (NSABB) was established in the US to provide recommendations to the federal government regarding topics such as the oversight of dual experiments and the communication of research findings. In 2007, it developed criteria for identifying research of concern that was meant for uptake by those receiving federal funding, and since then the US government has been deliberating how to implement the recommendations.

One noteworthy feature of the various risk–benefit reviews brought in is their conclusions: at least prior to late 2011, no manuscript or funding proposal[11] has ever been rejected on security grounds. Perhaps even more noteworthy with these review processes is the infrequency with which they have identified any items 'of concern' in the first place.[12]

In late 2011, the debate over the justification for oversight procedures was rekindled when American and Dutch researchers indicated they had mutated the H5N1 virus in a ferret model in such a way as to enable it to transmit readily between mammals. Subsequently the NSABB recommended that the researchers and the journals to which they submitted this work redact key details in order to prevent others from being able to recreate the experiment. The NSABB later reversed its decision, supporting the publication of revised versions of the original manuscripts. While the clamour and controversy associated with the communication and control of this set of experiments is still alive at the time of writing, what is perhaps most notable about this case is its exceptionality.

Certainly one of the reasons why risk–benefit review processes have almost never halted grant applications or manuscripts is the difficulty of determining risks associated with a single item of research against the backdrop of pre-existing work, by an unspecified user, and at an unknown time in the future. Given this experience, it seems much more sensible to ask what should be funded in the first place; for instance, whether biodefence work is really necessary, whether some forms of international funding promote human security by enhancing international collaboration and development or whether alternative research agendas can serve disease prevention.

Codes of conduct

Against the wide ranging and perhaps irresolvable questions associated with preventing the life sciences from becoming the death sciences, 'codes of conduct' have been repeatedly forwarded since 2001. Codes are commonplace in many professional contexts, though their aims vary tremendously. In relation to concerns about destructive applications of the life sciences, codes have often been portrayed as offering a means of self-regulation adept enough to keep pace with fast changing scientific developments.

An examination of the history of codes of conduct activities over the last decade indicates the tension-ridden aspects of the biosecurity initiatives that are the topic of this chapter. On the one hand, the introduction of codes that have a bearing on professional standards or behaviour has been limited: to the extent codes have been introduced, they have almost exclusively been advisory in nature and with limited practical implementation (Rappert, 2007). On the other hand, the discussion of the potential of codes and what should be in them has provided subject matter for furthering communication and collaboration between a wide range of organizations. In this sense, the important issue with evaluating code of conduct initiatives has not been that of whether they have worked or whether they could work, but instead asking what working means. Arguably codes have worked in the sense that the talk about them has served as a basis to enrol individuals and organizations into security discussions foreign to many (Rappert, 2009). Yet, this is not the function that has been sought from or attributed to them in policy decisions.

Education

After the events of 2001, professional organizations such as the World Medical Association, science advisory bodies such as the InterAcademy Panel and many national academies, the International Committee of the Red Cross, UN agencies, and countries such as Germany, Australia, Pakistan and Japan all stated the need for scientists to be knowledgeable about security dimensions of their work. Greater awareness of biosecurity issues through education has been promoted as part of building a culture of responsibility (see NSABB, 2008; National Security Council, 2009).

Such calls were buttressed by studies suggesting that the dual-use potential of research was a non-issue at a personal and professional level for many of those associated with the life sciences; that the understanding of safe laboratory practices needed improvement; and that there are low levels of formal training about matters of security as part of university curricula (see Rappert, 2010).

Despite the high-level accord about the importance of education by many governments and professional organizations, arguably what counts as appropriate education is still a matter of some dispute. This is so because calls to educate are inextricably bound up with the exercise of authority. Whether, for instance, the goal should be to transmit authoritative knowledge and values, or to nurture individuals' own reasoning, are matters on which disagreement is evident. When these are set on the international stage with all of its power asymmetries and histories, the question of what should be done by way of education becomes debateable (Bezuidenhout, 2012). In relation to the theme of this chapter, the question of whether biosecurity equates to concerns about bioterrorism, accidental releases or the threat of any type of disease is a major question for the content of any educational message.

The next 10 years

In light of the tension-ridden and shifting experience of the past 10 years, what might the next 10 years hold for the area of 'biosecurity' mapped in this chapter? While speculation about the future is always hazardous, we would posit three suggestions.

First, while the past decade has been characterized by a struggle for the understanding and recognition of biosecurity, the next one will be characterized by more technical and pragmatic matters pertaining to implementation and institutionalization. This dynamic is perhaps best exemplified by reference to the BWC. In 2001 and 2002, international diplomatic discussion fell into rancour with the failure to agree a verification protocol for the convention. Since that time the yearly meetings have sought to find common ground and define an agenda for discussion. With the growing recognition of biosecurity as a term, the introduction of new national policies and also the establishment of dedicated organizations and funding streams in many countries, previous attempts to find a shared language and agenda will increasingly give way to reporting on activities and building on initiatives.

Second, despite this movement towards implementation and institutionalization, the meaning of 'biosecurity' is likely to broaden out further and become more diffuse. There is nothing necessarily paradoxical in this simultaneous digging in and hollowing out. As national security-inspired concerns over biosecurity become integrated into other agendas, its meaning is likely to become ever more open to interpretation and all-encompassing.

Third, the future will be driven by events. The attention to and conceptualizations of biosecurity in the last decade owed much to specific events – the attacks of September 11th and the anthrax attacks that followed being the most prominent. The future is likely to be much the same. Further major terrorist attacks (or attacks of less magnitude with CBRN weapons) would almost certainly give momentum to biosecurity and shape how our first two suggestions play out. Additional contentious experiments might well bring to fruition research oversight procedures that have to date only been presented on paper.

Notes

1 Resolution 1540 (2004) adopted on 28 April 2004 at the 4956th meeting of the UN Security Council.
2 Remarks by the secretary-general to the Sixth Review Conference of the Biological Weapons Convention, Geneva, 20 November 2006.
3 A large grouping of primarily developing countries established in 1961 to take a middle course between the Western and Eastern blocs.
4 Security Council Committee established pursuant to resolution 1540, *Report of the Committee Established Pursuant to Security Council Resolution 1540*, 14 September 2011, S/2011/579, p. 2.
5 Ibid.
6 Fifty-fifth World Health Assembly resolution on 'Global public health response to natural occurrence, accidental release or deliberate use of biological and chemical agents or radionuclear material that affect health', 18 May 2002, WHA55.16.
7 Remarks on 'Global threats, global solutions' by Thomas Countryman, Assistant Secretary, Bureau of International Security and Nonproliferation, to the American Society for Microbiology, Biodefense and Emerging Diseases Research Meeting, Washington, DC, 27 February 2012.
8 Final Document of the Biological Weapons Convention Seventh Review Conference, Geneva, 5–22 December 2011, BWC/CONF.VII/7, p. 9.
9 Meaning those labs that are designated as able to work with agents that could cause serious or potentially lethal disease through inhalation.
10 These include: efforts to increase the virulence and transmissability of influenza viruses; the development of computer simulations that model the spread of disease; the creation of a chimera virus from components of an influenza virus and the West Nile virus; and the identification and characterization of antibiotic resistance to new antibiotics.
11 As in the reviews established by the UK Biotechnology and Biological Sciences Research Council, the UK Medical Research Council, the Wellcome Trust, the Center for Disease Control and the Southeast Center of Regional Excellence for Emerging Infectious Diseases and Biodefense.
12 See van Aken and Hunger (2009) and Rappert (2008).

References

9/11 Commission (22 July 2004) *The 9/11 Commission Report*, available at www.9-11com mission.gov.

van Aken, J. and Hunger, I. (2009) 'Biosecurity policies at international life science journals', *Biosecurity and Bioterrorism*, vol. 7, no. 1, pp. 61–72.

Barr, M. (2007) 'The importance of China as a biosecurity actor', in B. Rappert and C. Gould (eds) *Technology and Security*, London: Palgrave, pp. 121–132.

Bezuidenhout, L. (2012) 'Dual use issues in Africa', *Medicine, Conflict and Survival* (forthcoming).

Center for Arms Control and Non-proliferation (2008) *Federal Funding for Biological Weapons Prevention and Defense, Fiscal Years 2001 to 2008*, Washington, DC: Center for Arms Control and Non-proliferation.

Department of State (2001) Opening statement to the BWC Fifth Review Conference, delivered by John R. Bolton, Under-secretary of State for Arms Control and International Security, 19 November.

Douglas, M. and Wildavsky, A. B. (1982) *Risk and Culture: An Essay on the Selection of Technical and Environmental Dangers*, Berkeley, CA: University of California Press.

Enserink, M. and Kaiser, J. (2005) 'Has biodefense gone overboard?', *Science*, vol. 307, no. 5714, pp. 1396–1398.

Franco, C. and Sell, T. K. (2011) 'Federal agency biodefense funding, FY2011–FY2012', *Biosecurity and Bioterrorism*, vol. 9, no. 32, pp. 117–137.

Guillemin, J. (2005a) *Biological Weapons: From the Invention of State-Sponsored Programs to Contemporary Bioterrorism*, New York: Columbia University Press.

—— (2005b) 'Inventing bioterrorism: the political construction of civilian risk', in B. Hartmann, B. Subramaniam and C. Zerner (eds) *Making Threats: Biofears and Environmental Anxieties*, New York: Rowman & Littlefield, pp. 197–216.

Hayden, Erika Check (2011) 'Biodefence since 9/11: the price of protection', *Nature*, vol. 477, pp. 150–152.

Journal Editors and Authors Group (2003) *PNAS*, vol. 100, no. 4, p. 1464.

Kaiser, J. (2011) 'Taking stock of the biodefense boom', *Science*, vol. 333, pp. 1214–1215.

Klotz, L. and Sylvester, E. (2009) *Breeding Bioinsecurity*, Chicago, IL: University of Chicago Press.

Lakoff, Andrew and Collier, Stephen (2008) *Biosecurity Interventions: Global Health and Security in Question*, New York: Columbia University Press.

Leitenberg, M., Leonard, J. and Spertzel, R. (2004) 'Biodefense crossing the line', *Politics and the Life Sciences*, vol. 22, no. 2, pp. 1–2.

Lentzos, Filippa (2006) 'Rationality, risk and response: a research agenda for biosecurity', *BioSocieties*, vol. 1, no. 4, pp. 153–464.

Lentzos, Filippa and Rose, Nikolas (2009) 'Governing insecurity: contingency planning, protection, resilience', *Economy and Society*, vol. 38, no. 2, pp. 230–254.

Lentzos, Filippa and Silver, Pamela (2012) 'Synthesis of viral genomes', in Jonathan Tucker (ed.) *Innovation, Dual-use, and Security: Managing the Risk of Emerging Biological and Chemical Technologies*, Cambridge, MA: MIT Press.

National Research Council (2012) *Biosecurity Challenges of the Global Expansion of High-Containment Biological Laboratories*, Washington, DC: NRC.

National Security Council (2009) *National Strategy for Countering Biological Threats*, November, Washington, DC: NSC.

NSABB (National Science Advisory Board on Biosecurity) (2006) *Proposed Framework for the Oversight of Dual Use Life Sciences Research: Strategies for Minimizing the Potential Misuse of Research Information*, Washington, DC: NSABB.

—— (2008) *Strategic Plan for Outreach and Education On Dual Use Research Issues*, 10 December, p. 10, available at http://oba.od.nih.gov/biosecurity/PDF/FinalNSABBR eportonOutreachandEducationDec102008.pdf.

NSPD-33 (National Security Presidential Directive 33) (2004) *Biodefense for the 21st Century*, 28 April.

Rappert, B. (2007) *Biotechnology, Security and the Search for Limits: An Inquiry into Research and Methods*, London: Palgrave.

—— (2008) 'The benefits, risks, and threats of biotechnology', *Science and Public Policy*, vol. 35, no. 1, pp. 37–44.

—— (2009) *Experimental Secrets*, Plymouth, UK: University Press of America.

—— (ed.) (2010) *Education and Ethics in the Life Sciences*, Canberra, ACT: Australian National University E Press.

Rappert, B. and Gould, C. (eds) (2007) *Technology and Security*, London: Palgrave.

Revill, J. and Dando, M. (2007) 'The rise of biosecurity in international arms control', in B. Rappert and C. Gould (eds) *Technology and Security*, London: Palgrave, pp. 41–59.

Sunshine Project (2004) *Mandate for Failure: The State of Institutional Biosafety Committees in an Age of Biological Weapons Research*, Austin, TX: Sunshine Project.

Tucker, J. (2007) 'Strategies to prevent bioterrorism: biosecurity policies in the United States and Germany', in B. Rappert and C. Gould (eds) *Technology and Security*, London: Palgrave, pp. 213–237.

United States Government Accountability Office (2009) *High-Containment Laboratories: National Strategy for Oversight Is Needed*, Report to Congressional Requesters GAO-09-574 GAO, Washington, DC, available at www.gao.gov/new.items/d09574.pdf, accessed 7 January 2012.

Wheelis, M., Rózsa, L. and Dando, M. (eds) (2006) *Deadly Cultures: Biological Weapons since 1945*, Cambridge, MA: Harvard University Press.

The White House (2001) *President's Statement on Biological Weapons*, 1 November.

—— (2006) *The National Security Strategy of the United States of America*, 16 March.

—— (2007) *Homeland Security Presidential Directive/HSPD-21 on Public Health and Medical Preparedness*, 18 October.

WMD Center (2011) *Bio-Response Report Card*, October, available from www.wmdcenter.org.

WMD Commission (Commission on the Prevention of WMD Proliferation and Terrorism) (2008) *World at Risk: The Report of the Commission on the Prevention of WMD Proliferation and Terrorism*, 2 December, available at www.preventwmd.gov.

World Health Organization (May 2002) *Preparedness for the Deliberate Use of Biological Agents: A Rational Approach to the Unthinkable*, WHO/CDS/CSR/EPH/2002.16, Geneva: WHO, p. 2.

Wright, S. (2007) 'Terrorists and biological weapons: forging the linkage in the Clinton administration', *Politics and the Life Sciences*, vol. 25, no. 1–2, pp. 57–115.

Part IV

TRANSGRESSING BIOSECURITY

11

BIOSECURITY AND ECOLOGY

Beyond the nativism debate

Juliet J. Fall

Introduction: life's tendency to wander

Spotted knapweed (*Centaurea maculosa*) is a little plant with pretty purple flowers, native to Eastern Europe, yet hugely successful beyond this original range, particularly in North America. Spotted knapweed was accidentally introduced into North America in the 1890s, probably in alfalfa seed transported from Eastern Europe. It was first identified in Victoria, on the west coast of Canada, in 1893. It is assumed that soil carried on ships as ballast and unloaded in the port transported knapweed seed to this site (Mauer et al, 2001). Like all plants, it has a particular biography. Its very name reflects a classification: the term 'weed' typically denoting the unloved plants, growing in the wrong place, a purely cultural term devoid of botanical meaning. Botanists recently attempted to model the spread and potential future expansion of spotted knapweed (Broennimann and Guisan, 2008, p. 585). They explored what evolutionary and ecological factors influence the invasion process, testing whether plants evolve traits likely to increase their success in the new range (testing whether invaders are 'made') or whether functional determinants of communities or landscapes control invasiveness (invaders are 'born') because, surprisingly, the ecological conditions in both of these differed. Somehow, once they had moved halfway across the globe they appeared to prefer new ecological conditions. Simply assuming, as had been done up to then, that they would conserve ecological conditions similar to their native range was not enough (Broennimann and Guisan, 2008, p. 585).

Any traveller could have told you that travel broadens the mind, but this common-sense explanation might understandably not satisfy botanists. Instead, they suggested that one explanation is that certain plants, including spotted knapweeds, occasionally display different ploidal levels, that is to say that certain individuals have multiple copies of all their genes. Unless a taxonomist did extensive genetic analysis, two plants of differing ploidal levels would look exactly the same, and they would classify them as belonging to the same species. But this ploidal diversity is one possible explanation for the differing preferences between these globalized, mobile spotted knapweeds and their original sedentary cousins, challenging what we understand to be a single, particular species. These plants have used the globalized infrastructures that we have spread across the world, hopping on and off container ships from Europe to North America and back again, catching rides on lorries and spreading along roads, into marginal urban spaces. They have shown themselves to be out of bounds, out of place and out of control. Farmers curse them halfway across the globe. Throughout these debates, the question of where such plants came from originally (native range), before they started travelling, is a recurrent concern. To where do they *really* belong, what right do they have to settle in new places and are the new 'invaders' returning to Europe really fundamentally the same as the 'natives' they left behind? And, perhaps more importantly, to what extent are these ecological or political questions, and why might it matter?

In this chapter, and moving on from the example of the globally mobile and invasive knapweed, I explore how the idea of nativism structures both conservation policy and the public's sanctioned relationships with nature. This is a highly contested terrain, receiving critique and debate from a wide variety of natural scientists, social scientists, activists and stakeholders. It continues to be a difficult dialogue, with tempers flaring on all sides. Debates about the definition of what is 'natural', and about the separation of humans and nature, take a specific and meaningful form in the biosecurity context. This chapter therefore considers the language and definitions used to structure nativist concerns, the suitability of classification criteria, the underpinning science and the pragmatic justifications for nativist policies. Crucially, it discusses the discursive and political implications of the 'nativist paradigm', and the ideological assumptions and cultural motivations for nativist policies, including the degree to which the 'native good, aliens bad' discourse might be a barrier to all citizens' participation in environmental conservation, including ethnic minorities. Concrete examples draw from other research carried out in Switzerland where the issue of invasive species was specifically raised on the political agenda following the arrival of ragweed (*Ambrosia artemisiifolia*), a North American plant with human health impacts.[1] It ends with a discussion of what happens when policies are put into practice, indicating that on the ground categories are much less fixed than current academic debates might suggest.

Putting nature into boxes

It is sometimes difficult to remember that our current Western way of thinking about nature as 'biodiversity' is recent. It replaced previous valuations of nature as an example of the sublime creativity of a divine Creator for example and is a markedly different idea of nature from that prevalent in many other non-Western cultural traditions (Descola, 2005). The idea of biodiversity institutes a valuation of the degree of variation of life forms within a given species, an ecosystem or a biome. Crucially, it also involves awareness and lamenting of the decline of such diversity (Takacs, 1996). But it is not just a straightforward question of quantity, for, if it were simply about numbers, then adding one more species to what is living in a given place would surely be celebrated? Yet this is crucially not the case for exotic invasive species.

Not so long ago, however, adding chosen species to particular landscapes was specifically encouraged and institutionalized within acclimatization societies, particularly as part of state-sanctioned colonial projects. The first of these learned societies were founded in France and Britain in the middle of the nineteenth century, when moving species around the globe had fundamentally positive connotations. Both improving the supposedly defective colonial landscapes and rendering the metropolis exotic and cosmopolitan were noble aims embraced by many scientists (Crosby, 1986; Osborne, 2000; Smout, 2003). Lamarckian transformism, or a malleability of form and function, underpinned the acclimatization theories of colonial times, something that the spotted knapweed, mentioned at the beginning of this chapter, could have been thought to be displaying if we hadn't come up with other theories.[2] However, as Hall (2003, p. 8) writes,

> by 1900, exotic organisms were falling into disfavour, especially as ecological awareness began to expose mechanisms by which desirable native species succumbed to exotic competition. The cycles of fashion might also help to explain why exotic gloom began to eclipse exotic glory.

Yet, as many gardeners can tell you, planting species from elsewhere continues to have great appeal: those nurtured in back gardens specifically because they appear to bring a whiff of glamorous exoticism to the everyday. This love of exotic plants is not trivial: almost two-thirds (62.8 per cent) of the established plant species in Europe now listed as invasive were introduced intentionally for ornamental, horticultural or agricultural purposes (Keller et al., 2011; see also McNeill, 2003), often gaining strength by passing through the space of the garden where human selection and domestication, and a preference for exotic, colourful, vigorous, undemanding plants, has led to the creation of super-plants that have become feral (Jeanmonod and Lambelet, 2005, p. 14; Mack, 2001).

The idea of biodiversity nevertheless inexorably changed our relationship to these localized elsewheres, overturning not only how we think about place, nature and the environment, but also specifically how we identify who is responsible for defining and solving specific problems. The crucial role of conservation biologists in coining the term biodiversity as a concept uniquely suited to advocating action in the face of catastrophic decline is well known and documented (Takacs, 1996; Mauz and Granjou, 2010). Biodiversity centres on an accounting paradigm that involves thinking of nature as individualized species and specific assemblages, paradoxically reflecting both a carefully evolved order and a capacity for change. Yet, despite this apparent focus on change, it is the question of the order and permanence of nature that appears to be particularly prevalent in the popular imagination, and that receives much attention as it is translated into governance policies. Representations of nature as the Garden of Eden, with Nature viewed as a single entity reflecting a divine and perfect order (Macnaghten and Urry, 2000), may well be far from contemporary non-equilibrium biological models, but they still continue to influence many policy debates and mobilize for action.

Today, however, the ideal of improving nature and making it more harmonious, once the preserve of acclimatization societies, involves not the addition but the subtraction of selected species deemed unworthy. These invasive species seem to be rhetorically doubly perverse: not only are they spreading in new and unexpected ranges, they are also – like the spotted knapweed – crossing into spaces considered removed from nature: derelict train stations, industrial zones and other abandoned margins of human activities. Indeed, authors have argued that understanding what has been called the human preparation of landscape is a key factor in making sense of invasions (Robbins, 2004), as this happens when and where the invasiveness of certain species uniquely combines with the invadability of certain landscapes.

Classification: an agonized debate

There has been fierce debate in both social science and natural science journals about the specific terms of the debate around invasive species (Head and Atchison, 2008; Warren, 2007, 2008; Richardson et al., 2008; see Fall, 2011 and Fall and Matthey, 2011 for more details of these debates). This has been an agonized conversation, with tempers flaring and accusations of racism or xenophobia making interdisciplinary debate difficult (see Gröning and Wolschke-Bulmahn, 2003, about suggested links between native plant enthusiasts and Nazi history; and the strong response by Uekötter, 2007). These critiques of the terms and categories used have largely taken two paths, to some extent following disciplinary traditions: natural scientists worry that emotive categories (alien, invasive, native and so on) are scientifically inaccurate and counterproductive, and require refinement and streamlining in order to be more useful; while social scientists critique the fundamentally political nature and assumptions of such categories.

Many natural scientists accept that the rapid ascension in the public domain of the field of invasion biology owes a lot to the extensive use of adjectives such as 'invasive', 'alien', 'noxious' and 'exotic'(Colautti and MacIsaac, 2004, p. 135) that have immediate appeal. Broadly, and although not all authors agree, invasiveness generally refers to the behaviour of an organism, while alienness (and, conversely, nativeness) refers to its belonging in a certain place (Head and Muir, 2004). In critiques of such terms made by natural scientists, it is mainly confusion over terminology that is seen to have impeded progress in scientific theory (Colautti and MacIsaac, 2004, p. 135), since 'science progresses best when hypothesis, theories, and concepts are concisely stated and universally understood' (Colautti and MacIsaac, 2004, p. 139). That terms such as native and exotic are deeply historic, and highly charged, is well known (Hall, 2003; Olwig, 2003; Staszak, 2008) and increasingly acknowledged by invasion biologists. Colautti and

MacIsaac (2004) suggest that words like invasion and alien are in fact scientifically counterproductive, since subconscious associations with preconceived terms, particularly emotive ones, can lead to divergent interpretations and a confusion of concepts, clouding conceptualization of the processes they are meant to describe. This leads to 'widely divergent perceptions of the criteria for "invasive" species' (Colautti and MacIsaac, 2004, p. 135), 'lumping together of different phenomena, and the splitting of similar ones, which in turn makes generalization difficult or impossible' (Colautti and MacIsaac, 2004, p. 136). This, they argue, has led to deep divisions between invasion ecologists (Davis and Thompson, 2001; Davis et al., 2001). Suggestions have therefore been made to define terms unambiguously, based for instance on impact (Davis and Thompson, 2001) or on successions of ecological stages, in other words following ecological processes dependent on spatial scale (Colautti and MacIsaac, 2004). The latter posits that invasion is not the same as colonization, as other authors have argued. Instead, in this position, invasive species are somehow uniquely Other: 'we argue that the process of becoming nonindigenous is inherently different from the local spread characterized by native colonizers' (Colautti and MacIsaac, 2004, p. 137). Others, more simply, have noted that no species is inherently alien, but only with respect to a particular environment at a particular moment, and that such categories are of course not discrete, tightly defined and unambiguous terms, but cluster concepts with overlapping boundaries (Warren, 2007; Willis and Birks, 2006). As the example of spotted knapweed showed at the outset, not only is defining a native range not entirely straightforward, but neither is assuming that subsequently spreading plants will only adapt to conditions similar to it.

In response to what is viewed as problematic confusion over classification, others have argued that the problem isn't only scientific accuracy, calling instead for encouraging

> critical reflection on whether metaphors currently used to characterize these species may actually undermine conservation objectives …, [since] invasion biologists and conservation managers presumably (and perhaps unconsciously) rely on the rhetorical power of this language to generate action against these species, which are invisible to most people. Perhaps this approach has been successful, given the tremendous amount of attention this issue has received recently; nonetheless, these metaphors also pose a number of risks.
>
> (Larson, 2005, p. 495)

and noting that these metaphors invoke militaristic ways of thinking and that not only are they inconsistent with sustainable relations between humans and the natural world, but also that framing the problem as a war requires recognizing two opposing sides, which is paradoxical when their spread is inextricably entangled with human consumptive activities and global movement patterns (Larson, 2005, p. 496). In response to such concerns, less emotive terms have been suggested, and the terms neophyte (new plant) and neozoaire (new animal) have been suggested in French-speaking Switzerland, for example, but don't have the public visibility of previous terms (Klaus, 2002). Other authors have recommended a return to past cultural categories cast aside by the success of new terms focussing on place of origin: pest/non-pest and vermin, although the latter is equally historically loaded as it enjoyed great success in the Nazi era (Smout, 2003).

In practice, classifying plants as invasive on a national list is inevitably fraught, and more political than might be expected. In Switzerland, when this was carried out, the existing widely recognized definition suggested by the World Conservation Union (now IUCN) only referred to threats to biodiversity. Yet, faced with a growing ragweed invasion, the Swiss authorities were eager for economic and public health criteria to also be taken into account, something that an updated IUCN definition subsequently integrated. One of our interview partners (T2) recounts what happened:

> So, for certain species it was quite clear, for others it was a bit harder and especially when it came to the Watch List it was much harder to decide what we put on it and what we don't want. Where we draw the line. And also, at the beginning, we were a bit ambitious; we tried to have a Black List, a Grey List and a Watch List. This Grey List disappeared over time and then the species on that list were split between the two others.
>
> (T2)

These negotiations were necessarily very place-specific as different species created different problems in different places, and questions and criteria emerged collectively during discussions within the group of botanists. This political exercise of categorization, in which experts collectively define criteria, framing the problem as they go and building on inevitably partial knowledge, goes some way to show that, beyond the agonized debates about terms and categories, the actual practices of making invasive species into a coherent category depends on place-specific negotiated practices, rather than objective, aspatial rationality, as I discuss later on in the chapter.

Nativism: spatializing and ethnicizing the right to belong

Regardless of what terms are used, a clear nativist tendency runs through debates on invasive species: the idea of a discrepancy between the interests and rights of certain established inhabitants of an area or state as compared to claims of new-comers or immigrants. This might not be a problem in the natural sciences if, as is sometimes claimed, political and ecological domains were fundamentally different, with environmental concerns determined by value-free and science-led paradigms. Yet in the case of invasion ecology this assumption is revealed for the fallacy (or myth) that it is. Ecological policies are far from being a politics-free zone, since these reflect in multiple complex ways the underpinning social values of the societies that give rise to them (Robbins, 2004; Barker, 2010). Authors have argued that the continuing ambivalence toward the nature–society dichotomy and the 'longing for lost community purity … guides nativism aimed at both humans and non-humans alike' (O'Brien, 2006, p. 65). It is particularly ironic that much of the writing about invasive species is taking place in former colonies in which issues of 'indigenousness and belonging are discussed in contexts where settler human populations are still coming to terms with their own belonging' (Head and Muir, 2004, p. 203), by authors focussing largely on former European colonies (Barker, 2010; Crosby, 1986; Clark, 2002; O'Brien, 2006; Robbins, 2001, 2004). I would tentatively go further and suggest that because many of these countries – Australia, Canada, the United States, New Zealand and South Africa in particular – are English-speaking and the studies produced by local researchers are widely read, these forms of post-colonial guilt and anxiety about identity end up orienting ecological debates in ways that still need fully examining in contexts with very different ecological and social histories.

The link between states, nativism and nature is further paradoxically strengthened by our current way of thinking about nature as biodiversity. The ratification of the Convention on Biological Diversity instituted states as the official guardians of biodiversity, rather than making it a common heritage of humankind. Although this came about because of a concern related to intellectual property rights, and was not in any way a reflection of any specific or underlying nationalism, this *de facto* nationalization of biodiversity has a perverse consequence in the case of invasive species. Since state parties (i.e. each signatory country to the Convention) have to produce national lists of invasive species, this further entrenches decades-old ideas of direct links between the shape of the nation, nature and identity (Olwig, 2003, p. 72).

Thus what are presented as value-free tales instead tell of swarming, invading, foreign and out-of-control natures, and play on and to other fears, opportunistically rewriting the nation-state as the most pertinent scale of identity politics. In concrete terms, national black lists and watch lists select and make visible what are seen as the worst offenders. The question of scale and the identification of pertinent scales at which to define these ecological threats and possible governance policies to regulate movement become a key focus since the simple addition of local or national scenarios into 'global black lists' of invasive species, as listed for instance by the Global Invasive Species Database (GISD, 2005), is paradoxical since every species listed originally comes from somewhere (see Fall, in press). Yet this apparent inevitability of the nation as the scale of planning cannot make ecological sense: and the nation, as a socio-political construct, operates on a number of scales, leading some authors to suggest that 'in the European context, nation may be too small; in the Australian context, it may be too large' (Head and Muir, 2004, p. 199).

Beliefs inevitably find their way into conservation discussions, and the role of language and categorization in transporting values beyond the intent of any individual or group of speakers is well known, such as in the well-rehearsed foreigner-as-threat terminology (O'Brien, 2006, p. 67; see also Gould, 1997; Coates, 2003). Olwig (2003, p. 61) summarizes the problem clearly in writing that:

> the natural scientists who worry about the penetration of alien species often appear to be unaware of the parallels between their discourse and that of racists and national chauvinists. Few of these scientists would presumably wish to be classified as such. Yet racists and nationalists have been known to legitimate their arguments by drawing parallels between the arguments of scientists concerning ecological imperialism and the supposed threat of foreign species, on the one hand, and, on the other, the perceived threat of foreign races and cultures to the native populations of their countries.

Likewise, Coates notes that demeaning, defaming and othering specific groups of people through associations with creatures not only reinforces social and racial privileges by lending them the weight of natural authority, but 'it also facilitates beastly behaviour toward the animalized and the naturalized' (Coates, 2003, p. 135).

This strand of critique therefore takes a different, more fundamental route. It builds on the concern about the prevalence of metaphors in the natural and ecological sciences (Larson, 2005). While metaphors are ubiquitous in science, 'simplicity and intuitive appeal are also the main reason why scientific language has never succeeded in "cleansing" itself from metaphorical "impurities", despite several attempts to do so' (Chew and Laubichler, 2003, p. 52). Interpreting natural phenomena in human terms is, however, 'a two-edged sword, generating knowledge as well as opening the door to troubling misunderstandings' (Chew and Laubichler, 2003, p. 52). Metaphors thus introduce a fundamental trade-off between the generation of novel insights in science and the possibility of dangerous or even deadly misappropriation (Chew and Laubichler, 2003; see also Rémy and Beck, 2008). This is what Cresswell warns about when he writes about biological morality, in which metaphors are not just theoretically inappropriate, but can also have 'serious consequences on people's lives' (Cresswell, 1997, p. 336).

Yet metaphors, though at times deadly, are useful in that they allow us to build on our experience when we extend familiar relationships to unfamiliar contexts, helping new ideas to spread. In the natural and ecological sciences, where much is inferred rather than directly observed, metaphors can make the difference between comprehension and confusion, helping to get a message across (Chew and Laubichler, 2003, p. 53). The frequent use of war-like ones such as invasive species or natural enemies nevertheless appears to imply that such categories exist in nature, assigning a normative dimension to the metaphors. While such categories can only be idealized abstractions, their existence is reinforced by the metaphorical language of scientists. They end up becoming concrete objects (Chew and

Laubichler, 2003, p. 53). That environmental policies now rely on similar framings of perpetual war and permanent vigilance, with terms such as 'sleeper species' mirroring those used to describe terrorist cells – and leading to subsequent networks of surveillance and control that mirror those watching stigmatized human populations (EEA, 2010) – is further indication of how the framing of ecological concerns builds upon social and political contexts.

Challenging the permanence of states: anxieties about globalization

While fears of invasion seem to feed off anxieties about globalization and global change – suggested by the strategic coining of the term global swarming in reference to contemporary and established fears of climate change – they are nothing new. Anxiety and fear of future scenarios of rapid change has always been at the heart of invasion biology. Charles Elton, largely recognized as the founder of the field, was writing his key text in 1958 at a time of increased global anxiety:

> Nowadays we live in a very explosive world, and while we may not know where or when the next outburst may be, we might hope to find ways of stopping it or at any rate of damping down its force. It is not just nuclear bombs or wars that threaten us, though these rank very high on the list at the moment: there are other sorts of explosions, and this book is about ecological explosions.

> (Elton, 1958, p. 15)

At the height of the Cold War, he did not shy away from using emotive vocabulary to make his case, and he explained that population explosions – plants, animals, but also pathogens – included 'those that occur because a foreign species successfully invades another country … [bringing about] terrific dislocations in nature' (Elton, 1958, p. 18). These formulations of the problem around states ('another country') and on the idea that invasive species were somehow anti-natural ('dislocations') were extremely influential and largely continue to frame research and policies today. Equally important, the field of invasion biology – like, later on, the term 'biodiversity' – were coined specifically to combine scientific description and analysis with awareness-raising and calls to action in the face of urgent, and presumably catastrophic, changes.

The intimacy between social metaphors and claims about exotic nature means that discourses play on numerous feelings of insecurity and fear of difference. Clark (2002), for instance, further provides pathways for exploring the rich vein of symbolic association of social dia*spora* and cosmopolitanism with bad seeds (dia*spores*), weeds and vermin, linking up with the literature on the risk society of Beck (1999). The right to belong indeed echoes many other contemporary fears, about human migration and threatening foreigners, making the question of a 'war on invasives' all the more charged emotionally as it connects to the right to belong for various categories of persons in particular places. Yet, for many of us, belonging to many places at once and quickly feeling at home in new places has become commonplace. Despite these incredible changes in our social and political worlds, it might therefore seem paradoxical that we continue to worry about what really belongs where when we think of plants and animals. Thus the effect of this framing of the problem must be acknowledged as problematic in many multicultural societies where authorities profess a desire to see all segments of society reap the benefits of environmental protection and make use of existing green spaces.

Eager to test how ordinary citizens perceive official and lay discourses on invasion, including documents using what we identified as emotive or highly charged language, we organized five focus groups of citizens in Geneva within which we discussed a wide variety of documents, from newspaper articles to official information leaflets on specific species.[3] To our surprise, repeated strategies of distancing emerged within the groups, constantly challenging the pertinence of the category itself of 'invasive species'. Individuals made repeated connections with other harmful but native species to dispute the need to take action, such as stinging nettles (*Urtica dioica*), poisonous mushrooms, or native invasive plants. Many also suggested that there must be an underlying naturalness to such processes, or to the presence of particular species, and disputed how the documents presented specific species as somehow unnatural. These focus groups also highlighted how communicating with diverse publics on invasive species is in any case notoriously difficult since it involves a number of paradigm shifts. Where formally environmental and conservation groups tried to convince the public that less intervention and more pristine nature was better, and that all forms of life were intrinsically valuable, some of these same groups now call for special dispensations to use banned herbicides in nature reserves (Nicolas Delabays, 2010, pers. comm.) and promote heavy-handed mechanical intervention with diggers in natural areas. It should, however, be noted that there are important variations in such practices (Kowarik, 2003, 2011; Lachmund, 2004), and that more cosmopolitan approaches to welcoming invasive species are beginning to return.

Conclusion: stepping out of boxes

This chapter has reviewed and commented on many of the debates surrounding questions of nativeness, debates that at times appear to be going round in circles. However, in our Swiss case study, one of the most interesting things to emerge from interviews with practitioners was the numerous negotiations and transgressions around what were presented as established expert categories and consequent environmental policies. Many of the individuals interviewed were employed in public bodies and were working either in city parks or in nature reserves. While many demonstrated clear loyalty to the cause of fighting invasive species, they also repeatedly justified not doing so in a number of cases. Some argued for instance that particular species were not actually on a black list at all and that therefore no action was required;[4] that others were not really a problem in their opinion and that this trumped official policy; that there wasn't sufficient time or interest to take action, or that it was too late to do anything; or even that aesthetic or external political reasons made removal impossible. A city park employee in Geneva known as 'T6', for example, mentioned a number of pragmatic exceptions, including the following:

> We make an exception in the case of the pawlonia [*Paulownia tomentosa*] … because this is clearly a tree with a certain decorative value, and we haven't noticed that there was an important spreading of pawlonias, unlike the ailanthus tree [*Ailanthus altissima*] that is all over the place. So we allowed trees that had been cut down to be replaced.
>
> (T6)

> Or else, for example, if we take the Route des Acacias, we have lines of acacia trees [*Robinia pseudoacacia*]. An acacia dies, in that whole line, and well, we are not going to plant an oak tree [*Quercus*], or a chestnut [*Aesculus hippocastanum*] in the middle of a line of acacias, it's a question of how we view the landscape.
>
> (T6)

> After that, if we take out all the laurel [*Kalmia latifolia*] hedges in the green spaces [urban parks], I cannot tell you how much work that would be, it's quite simply enormous.
>
> (T6)

In a sense, the examination of such practices goes some way towards indicating that, beyond the agonized and somewhat circular debates surrounding the definition of categories, the actual pertinence of the homogeneous category of 'invasive species' is challenged when it comes to implementation and practical measures – or when it is relocalized, to use Miller's (2004) productive term – just as it was by the participants in our focus groups. The issue of invasive species in Switzerland was constituted as a singular problem and thereby achieved wide political recognition through the strategic alliance of natural scientists, environmental groups and health professionals concerned about the public health impacts of increasing numbers of ragweed. Yet this coherence of the category of 'invasive species' is challenged by the diminishing visibility of this flagship species: this specific problem appears under control, and the expected catastrophic rise in respiratory allergies did not take place. This is further strengthened by the increasing recognition that the individual biographies and behaviours of each species require extremely different measures to control them, or to accept them as new inhabitants while hoping for a return to some sort of equilibrium in the future. Focussing on practices, rather than on discourses, certainly offers further interesting paths to understanding, and challenging, the categories we craft to make sense of the living world around us.

Notes

1 This draws from a research project at the University of Geneva funded by the Fondation Boninchi, carried out from January 2010 to June 2011. Some direct quotes from interviews are translated by me from the original French, while one interview was carried out in English. Laurent Matthey, Irène Hirt and Marion Ernwein were involved at various crucial stages, for which I am immensely grateful. Anonymity was granted to some interviewees when requested, and transcriptions were numbered T1 to T16.
2 See Bernardina (2000) on myths of metamorphosis and invasive species.
3 As this was an exploratory piece of research, groups were constituted of 6–9 people in a rather *ad hoc* manner, partially through targeted advertisements inviting participants 'to discuss an environmental issue'. More by chance than by design, these reflected the diversity of the local population, in terms of nationality and place of origin. It must be noted, however, that the question of minority or ethnic groups' reception of discourses on invasive species was not a specific research topic.
4 The coexistence of and confusion over the presence of multiple, shifting black lists with various differing legal statuses and territorial extent made such an argument all the more plausible in the case of Switzerland. In the canton of Geneva, for instance, at least three different lists of species coexist and are referred to, with surprisingly large discrepancies between them.

References

Barker, K. (2010) 'Biosecure citizenship: politicising symbiotic associations and the construction of biological threat', *Transactions of the Institute of British Geographers*, vol. 35, no. 3: 350–63.

Beck, U. (1999) *World Risk Society*, Polity Press, Cambridge.

Bernardina, S. D. (2000) '"Algues tueuses" et autres fléaux. Pour une anthropologie de l'imaginaire écologique en milieu marin: le cas de *Caulerpa taxifolia*', *La Ricerca Folklorica*, vol. 42: 43–55.

Broennimann, O. and Guisan, A. (2008) 'Predicting current and future biological invasions: both native and invaded ranges matter', *Biological Letters*, vol. 23, no. 4(5): 585–89.

Chew, M. K. and Laubichler, M. D. (2003) 'Natural enemies – metaphor or misconception?', *Science*, vol. 301, no. 5629: 52–53.

Clark, N. (2002) 'The demon-seed: bioinvasion as the unsettling of environmental cosmopolitanism', *Theory Culture Society*, vol. 19, no. 1–2: 101–25.

Coates, P. (2003) 'Editorial postscript: the naming of strangers in the landscape', *Landscape Research*, vol. 28, no. 1: 131–37.

Colautti, R. J. and MacIsaac, H. J. (2004) 'A neutral terminology to define "invasive" species', *Diversity and Distributions*, vol. 10: 135–41.

Cresswell, T. (1997) 'Weeds, plagues, and bodily secretions: a geographical interpretation of metaphors of displacement', *Annals of the Association of American Geographers*, vol. 87, no. 2: 330–45.

Crosby, A. (1986) *Ecological Imperialism: The Biological Expansion of Europe 900–1900*, Cambridge University Press, Cambridge.

Davis, M. A. and Thompson, K. (2001) 'Invasion terminology: should ecologists define their terms differently than others? No, not if we want to be of any help', *Bulletin of the Ecological Society of America*, vol. 82, no. 3, 206.

Davis, M. A., Thompson, K. and Grime, J. P. (2001) 'Charles S. Elton and the dissociation of invasion ecology from the rest of ecology', *Diversity and Distributions*, vol. 7, no. 1/2: 97–102.

Descola, P. (2005) *Par-delà nature et culture*, Bibliothèque des sciences humaines, Paris, Gallimard.

Elton, C. (1958, reprinted 2000) *The Ecology of Invasions by Animals and Plants*, Methuen, London. Reprinted by the University of Chicago Press, Chicago.

European Environment Agency (EEA) (2010) 'Towards an early warning and information system for invasive alien species (IAS) in Europe', Technical Report no. 5, Office for Official Publications of the European Union, Luxembourg.

Fall, J. J. (in press) 'Governing mobile species in a climate-changed world', in J. Stripple and H. Bulkeley (eds) *Governing the Global Climate: Rationality, Practice and Power*, Chicago University Press, Chicago (draft submitted March 2012).

——— (2011) 'Invasions étranges, invasions étrangères, ou quand cygnes et écureuils bouleversent les frontières', in I. Dubied, A. Gerber, D. Fall and J. J. Fall (eds) *Aux frontières de l'animal: mises en scènes et reflexivités*, Collection Travaux de sciences socials, Droz, Geneva/Paris.

Fall, J. J. and Matthey, L. (2011) 'De plantes dignes et d'invasions barbares: les sociétés au miroir du vegetal', *VertigO – la Revue Électronique en Sciences de l'Environnement*. Débats et Perspectives, online 27 September. http://vertigo.revues.org/11046; DOI: 10.4000/vertigo.11046.

Global Invasive Species Database (GISD) (2005) '100 of the world's worst invasive alien species'. www.issg.org/database/species/search.asp?st=100ss, 10 September 2012.

Gould, S. J. (1997) 'An evolutionary perspective on strengths, fallacies and confusions in the concept of native plants', in J. Wolschke-Bulmahn (ed.) *Nature and Ideology: Natural Garden Design in the Twentieth Century* (pp. 11–19), Dumbarton Oaks Research Library and Collection, Washington, DC.

Gröning, G. and Wolschke-Bulmahn, J. (2003) 'The native plant enthusiasm: ecological panacea or xenophobia?', *Landscape Research*, vol. 28, no. 1: 75–88.

Hall, M. (2003) 'Editorial: the native, naturalized and exotic – plants and animals in human history', *Landscape Research*, vol. 1: 5–9.

Head, L. and Atchison, J. (2008) 'Cultural ecology: emerging human-plant geographies', *Progress in Human Geography*, vol. 33, no. 9: 1–10.

Head, L. and Muir, P. (2004) 'Nativeness, invasiveness, and nation in Australian plants', *Geographical Review*, vol. 94, no. 2: 199–217.

Jeanmonod, D. and Lambelet, C. (2005) *Envahisseurs! Plantes exotiques envahissantes: en savoir plus pour comprendre et pour agir*, Conservatoire et Jardin botanique de la Ville de Genève, Geneva.

Keller, R. P., Geist, J., Jeschke, J. M. and Kühn, I. (2011) 'Invasive species in Europe: ecology, status, and policy', *Environmental Sciences Europe*, vol. 23: 1–17.

Klaus, G. (2002) 'Nous devons communiquer la liste noire avec precaution', *Hotspot*, vol. 5 (Special issue on 'Biodiversité et espèces invasives: dialogue entre recherche et pratique'), pp. 12–13.

Kowarik, I. (2011) 'Novel urban ecosystems, biodiversity, and conservation', *Environmental Pollution*, vol. 159: 1974–83.

——— (2003) 'Human agency in biological invasions: secondary releases foster naturalisation and population expansion of alien plant species', *Biological Invasions*, vol. 5, no. 4: 293–312.

Lachmund, J. (2004) 'Knowing the urban wasteland: ecological expertise as local process', in S. Jasanoff and M. Long Martello (eds) *Earthly Politics: Local and Global in Environmental Governance* (pp. 241–62), MIT Press, Cambridge, MA and London.

Larson, B. (2005) 'The war of the roses: demilitarizing invasion biology', *Frontiers in Ecology*, vol. 3, no. 9: 495–500.

Mack, R. N. (2001) 'Motivations and consequences of the human dispersal of plants', in J. A. McNeely (ed.) *The Great Reshuffling: Human Dimensions in Invasive Alien Species* (pp. 23–34), International Union for the Conservation of Nature, Gland, Switzerland and Cambridge.

Macnaghten, P. and Urry, J. (2000) 'Bodies of nature: introduction', *Body and Society*, vol. 6, no. 1: 1–11.

McNeill, J. R. (2003) 'Europe's place in the global history of biological exchange', *Landscape Research*, vol. 28, no. 1: 33–39.

Mauer, T. M. J., Russo, M. J. and Evans, M. 2001 '*Element stewardship abstract for* Centaurea maculosa *(spotted knapweed)*', The Nature Conservancy, Arlington, VI, available at http://tncweeds.ucdavis.edu/esadocs/documnts/centmac.rtf, accessed November 2010.

Mauz, I. and Granjou, C. (2010) 'La construction de la biodiversité comme problème politique et scientifique, premiers résultats d'une enquête en cours', *Sciences Eaux et Territoires*, vol. 3: 10–13.

Miller, C. A. (2004) 'Resisting empire: globalism, relocalization, and the politics of knowledge', in Sheila Jasanoff and Marybeth Long Martello (eds) *Earthly Politics: Local and Global in Environmental Governance* (pp. 81–102), MIT Press, Cambridge, MA and London.

O'Brien, W. (2006) 'Exotic invasions, nativism and ecological restoration: on the persistence of a contentious debate', *Ethics, Place and Environment*, vol. 9, no. 1: 63–77.

Olwig, K. R. (2003) 'Natives and aliens in the national landscape', *Landscape Research*, vol. 28, no. 1: 61–74.

Osborne, M. A. (2000) 'Acclimatizing the world: a history of the paradigmatic colonial science', *Osiris*, vol. 2, no. 15 ('Nature and empire: science and the colonial enterprise'): 135–51.

Rémy, E. and Beck, C. (2008) 'Allochtone, autochtone, invasif: catégorisations animales et perception d'autrui', *Politix*, vol. 21, no. 82: 193–209.

Richardson, D. M., Pysek, P., Simberloff, D., Rejmanek, M. and Mader, A. D. (2008) 'Biological invasions – the widening debate: a response to Charles Warren', *Progress in Human Geography*, vol. 32: 295–98.

Robbins, P. (2004) 'Comparing invasive networks: cultural and political biographies of invasive species', *Geographical Review*, vol. 94, no. 2: 139–56.

—— (2001) 'Tracking invasive land covers in India, or why our landscapes have never been modern', *Annals of the Association of American Geographers*, vol. 91, no. 4: 637–59.

Smout, T. C. (2003) 'The alien species in 20th-century Britain: constructing a new vermin', *Landscape Research*, vol. 28, no. 1: 11–20.

Staszak, J.-F. (2008) 'Qu'est-ce que l'exotisme?', *Le Globe*, vol. 148: 7–30.

Takacs, D. (1996) *The Idea of Biodiversity: Philosophies of Paradise*, Johns Hopkins University Press, Baltimore, MD.

Uekötter, F. (2007) 'Native plants: a Nazi obsession?', *Landscape Research*, vol. 32, no. 3: 379–83.

Warren, C. R. (2008) 'Alien concepts: a response to Richardson et al.', *Progress in Human Geography*, vol. 32, no. 2: 299–300.

—— (2007) 'Perspectives on the "alien" versus "native" species debate: a critique of concepts, language and practice', *Progress in Human Geography*, vol. 31, no. 4: 427–46.

Willis, K. J. and Birks, H. J. B. (2006) 'What is natural? The need for long-term perspective in biodiversity conservation', *Science*, vol. 314, no. 5803: 1261.

12

INTRODUCING ALIENS, REINTRODUCING NATIVES

A conflict of interest for biosecurity?

Henry Buller

Introduction

This chapter explores the tensions between wild and managed natures, between biodiversity as dynamic natural variability and biosecurity as an intentional strategy of constraint. At one level, biodiversity and biosecurity might be seen as competing biopolitical paradigms (Buller, 2008), or 'competing modes of biopolitics' (Lorimer and Driessen, 2011), where the traditional separation between the 'natural' and the 'artificial' or 'cultural' is blurred and where technology and ecology, science and politics, intertwine to ultimately reinvent nature (Lemke, 2011).

For many, biodiversity is seen as inherently challenging to biosecurity, whether it be through the impact of 'non-native' or reintroduced species on the security of indigenous wildlife or through the spread of zoonotic disease and infection into secure systems of production. Similarly, biosecurity can be a threat to biodiversity, limiting and constraining natural adaptations and responses as well as interfering with often culturally cherished notions of 'wild' and 'natural' ecologies. At another level, however, biosecurity and biodiversity increasingly operate in parallel. 'A successful outcome for biosecurity is a successful outcome for biodiversity' claims the New Zealand Biodiversity Strategy (New Zealand Biodiversity, undated) advocating strict biosecurity measures to protect native biodiversity. Conserving natural variety, and its potential, in seed banks and protected areas, is seen as a long-term strategy for the future security of human populations and such conservation often relies upon a degree of biosecurity. Increasingly, the two are intertwined through species reintroduction programmes and the planned establishment of recombinant ecologies, where alien and indigenous species are intentionally brought together often to promote natural regrowth.

Although the aesthetic, biological and material constancy of wild nature is often seen as a much needed corollary to the frenzy and intensity of modern life, growing concern for biosecurity signals a shift in the long-negotiated and delicate balance between nature protection and production, and indeed the very parameters for such a distinction. The existing rationale for bio-protection is challenged as the threat of change grows. The exuberance and abundance of nature is increasingly incompatible with the macro- and micro-managed bio-technological spaces and processes of contemporary human endeavour. The function of protection shifts as natural and semi-natural spaces are redefined, no longer as refuges from change, but rather instruments in the management of change and security. Biodiversity conservation and the protection and maintenance of 'natural wild ecosystems' become highly significant objectives for new reasons (Locke and Mackey, 2009), which leads to the following questions:

- What weight to give the pull of the past against the fecundity of the present?
- Who will be the spokespeople and for what interests?
- How might the relative uncertainties (of time, of space, of cause and effect) be accommodated and what knowledges will be brought to bear?
- What value is to be found in those 'wild natures' that are increasingly seen as potentially destructive and uncertain when held against the highly managed and mediated natures of human productivity?
- What certainties of security, quality, health, spirituality and 'goodness' does 'wild' nature offer us if nature itself is revealed as fickle and changing?
- How might an aesthetic or a politics of stability and security be reconciled to a reinvigorated wild?

Hence, my concern in this chapter is with the (re)constructed biodiversity of re-wilding, restoration and reintroduction programmes: less the 'oncology' of a distant biology (Padel, 2005) than the 're-animation' of a more normative naturality.

Approaches to re-wilding

In February 2012, the Australian biologist David Bowman proposed the introduction of elephants into the Australian outback to reduce the extent and spread of non-native Gamba grass (*Andropogon gayanus*), which was brought in as a food source for farmed herbivores in the 1930s, and whose rampant growth is now having a highly detrimental effect on local flora and fauna (Bowman, 2012). Elephants, like foxes, cats, rabbits and cane toads before them, are the latest in a series of 'alien species' brought into Australia as an 'ecological tool' in the management of that nation's biosecurity.

Taking this perhaps a stage further, a group of American ecologists (Donlan et al., 2005a) has recently suggested a strategy of what they call 'Pleistocene re-wilding', which they define as:

> a series of carefully managed ecosystem manipulations using closely related species as proxies for extinct large vertebrates, and would

change the underlying premise of conservation biology from managing extinction to actively restoring natural processes.

(2005a, p. 913)

This would entail the importation of key 'Old World' mega-fauna, vaguely representative of the Pleistocene era, such as the Bolson tortoise (*Gopherus flavomarginatus*), feral horse species, African cheetahs (*Acinonyx jubatus*), lions and elephants, to the Great Plains and Midwest. Pointing out that around 77,000 large mammals (including Asian and African ungulates, cheetahs and kangaroos) already roam free on Texas ranches, Donlan et al. (2005a) argued that such re-wilding offers an 'optimistic alternative' to the current irrevocability of biodiversity loss.

At a less dramatic scale, in the UK, Natural England in association with the Otter Trust embarked on a reintroduction programme of the European otter (*Lutra lutra*) to the British countryside in 1983. Otters were all but extinct in British rivers, largely due to water pollution from pesticide residues. The otter population has been augmented by the planned reintroduction of 166 captive-bred individuals between 1983 and 1999, principally in the rivers of East Anglia, contributing to a significant increase in otter numbers at 56 per cent of observed sites (Natural England, 2010). However, other factors, such as improved water quality, declining use of pesticides and increased fish stocks are also major factors in this recovery. In this instance, re-wilding has been enrolled into a wider project of countryside restoration. The otter plays less the role of ecosystem captain than key indicator species for a romantic and culturally enriched rural biology, generous to otters but far less so to foxes.

American non-metropolitan space is not so easily packaged. If the British rural aesthetic is firmly embedded in a romantic Arcadian sensibility (Bunce, 1994), the American 'wild' reflects more recent cultural mediation (Wilson, 1992) and anxieties (Davis, 1998). The city of Chicago has, within its metropolitan area, a large number of designated wild land preserves which, over the course of the last 100 or so years, have not only experienced unmanaged forest regrowth and recolonization by a number of animal species, but have also become popular outdoor recreational sites. The mid-1990s saw the initiation of a series of ambitious, state-funded restoration projects aimed at returning parts of some of these reserves to their original, pre-settlement condition. For many, this would be tallgrass grassland and oak savannah (Mendelson et al., 1992). However, to achieve that ecologically authentic former status, woodlands had to be destroyed, wild deer either captured or shot and selective herbicides employed to prevent 'natural' regrowth until the restored prairie ecology had been fully re-established (Siewens, 1998). The unexpected public opposition to these restoration projects, which became known as the 'Chicago Restoration Controversy' (e.g. Gobster and Hull, 2000), led to a reappraisal of the schemes, in particular the way the schemes were managed. But, perhaps more significantly it demonstrated that 're-wild' nature in an urban or quasi-urban setting means something very different to 're-wild' nature elsewhere (Gobster, 1997).

These four brief examples reveal different rationales for the practice of 're-wilding' as a biopolitics of: security, conservation, restoration or manipulation. Yet more substantive differences can be found in the broader conception of re-wilding as a strategy of intervention.

Re-wilding is an elusive term. The British Ecological Society (BES) defines it as 'a specific case of landscape scale conservation that is defined as the "conservation of sites using only, or largely, natural processes"; using the relaxation of human management to return the site to a presumed previous state' (BES, 2009). Though it has received growing attention from UK policy makers and scholars (Sutherland et al., 2009), the term itself is contested. 'Wilding seems alien to the UK's rather cosy notion of nature', states the BES (2007), 'we simply don't do wild!' (BES, 2007, unpaginated). Beyond the UK, however, and at an altogether bigger scale, a more brazen re-wilding is heralded by its leading contemporary proponents as 'the scientific argument for restoring big wilderness based on the regulatory roles of large predators' (Soulé and Noss, 1998, p. 5). Presented by Soulé and Noss as the 'fourth current' in the modern conservation movement, following 'monumental-ism', 'biological conservation' and 'island biogeography', re-wilding, as advanced by Foreman (2004), draws upon three core principles: first, the key role played by top predators and key dominant species in maintaining the structure and resilience of ecosystems; second, the importance of protecting large areas of land for the movement of such animals; and third, the role of connectivity to allow movement between core protected areas. Given that such large species are either: (i) extinct from target areas for restoration, (ii) never dwelt there, or (iii) occur at such low densities that they have had little overall ecological function, re-wilding implicitly involves the translocation from other sites and introduction into places to which they are, to a greater or lesser extent, alien. This is re-wilding on a significant, almost continental, scale involving 'self regulating land commu-nities' (Soulé and Noss, 1998, p. 6). Yet, it is only one, albeit perhaps the most radical, interpretation of the concept of re-wilding. At least three other conceptions might also be identified: restoration ecology, native species reintroduction and de-domestication.

Restoration ecology has a slightly longer history than contemporary forms of re-wilding. Indeed, Cairns (2002, p. 16) identified the restoration of the Thames Estuary in the late nineteenth century as one of the first instances. Described by Cairns and Heckman (1996) as an emerging synthesis of ecological theory and concern about human impact on the natural world, restoration ecology, like re-wilding, makes claims to be both a 'science' and a 'practice': at one and the same time 'goal-oriented' and 'process-oriented' (Cairns and Heckman, 1996, p. 169), a mixture in short of historicism and futurism. Restoration ecology or ecological restoration has been formally defined as 'the return of an ecosystem to a close approximation of its condition prior to disturbance' (US National Research Council 1992 quoted in Bradshaw, 2002, p. 5) and, later, as 'the process of assisting the recovery and management of ecological integrity' (Society for Ecological Restoration, 1996, quoted in Bradshaw, 2002, p. 5). On a smaller scale than

Foreman's (2004) continental re-wilding, the focus here is the distinctive ecosystem unit, the boundaries of which can be as biological and 'natural' as they are fiscal, legal or economic. Restoration operates within varying cultural ideas of environmental degradation, leading to often subtle variations in style and objective. Hall (2000), for example, notes that, while Americans seek principally to restore wilderness, Italians have preferred the repair of gardens and managed landscapes. However, although we might take the above definitions as an invitation to return ecosystems to their formerly 'wild' state, almost every word here opens the door to debate and potential controversy (Davis, 2000). How far back should one go to achieve an ecosystem's 'condition prior to disturbance' (to the Pleistocene?); are only anthropogenic disturbances to be considered?; how much 'assistance' is legitimate and to what degree can the integrity of such restored systems be validated? In defence, Choi (2007) argues that we need to admit the significant social and biological limitations of past-facing and nostalgic restoration projects and, instead, promote a 'future-oriented' restoration whose goals are both sustainable within changing climatic and human socio-economic contexts and dynamic within the context of adaptive ecosystem functionality.

While restoration ecology is concerned with restoring (or rehabilitating) degraded ecosystems through multiple perspectives (e.g. Ehrenfeld and Toth, 1997) and techniques (Dobson et al., 1997), species reintroduction, as an intentional form of re-wilding, is more specifically targeted, usually, on a single species or animal, bird or plant. Like restoration ecology, this has a long heritage and, in recent years, has achieved a greater degree of legitimation as a practice recognized and encouraged by such international bodies as the International Union for the Conservation of Nature (IUCN) and the World Wildlife Fund (WWF). There are many examples of species reintroduction programmes across the world from tigers in Kazakhstan and red kites (*Milvus milvus*) in the UK to the Chiricahua leopard frog (*Rana chiricahuensis*) in Arizona and the African spurred tortoise (*Geochelone sulcata*) in Senegal (Soorae, 2011). The key lies in the 're' of reintroduction. As the IUCN specify, reintroduction is 'an attempt to establish a species in an area which was *once part of its historical range*, but from which it has been extirpated or become extinct' (IUCN, 1998, p. 6, emphasis added). Hence, while animals may be translocated from one site to another, it is generally to areas to which they are not wholly alien and out of place, even though considerable periods of time might have passed, and substantial ecological change occurred, since they were actually present in significant numbers. The more contentious boundaries of species reintroduction often fall, first, around issues of intentionality and the technologies of interventionism and, second, around concerns for the impact of reintroduced species on local wildlife and, in some instances, human social and economic practice. Moreover, the translocation of Persian leopards (*Panthera pardus ciscaucasica*) from donor zoos around the world to the Russian Caucuses under WWF-Russia's re-wilding programme is a very different form of reintroduction from the widely anticipated, though arguably 'natural', northbound migration of wolves from the northern Italian Alps into southeastern France.

In all of the above approaches to re-wilding, the principal dynamics are first spatial, involving the controlled translocation of species and the (re)establishment of boundaries. Second, they are temporal (Hall, 2009; Lowenthal, 2009), artificially speeding up time-deepened natural process through such techniques as 'nurse' and 'vector' species, simultaneous rather than sequential reintroductions, and so on. What is being re-wilded (or de-anthropogenized) are places, sites and ecosystems. The final approach to re-wilding or 'de-domestication' is somewhat different. Here, the targets include individual animals (for the most part). Through successive 'back breeding' techniques, or through rigorous protection by traditional breeding and genetic manipulation, formerly domesticated animals and breeds are rendered 'wild' anew and introduced (or reintroduced) into restored ecologies with the intention of re-establishing or regenerating wild(er) places, behaviours and ultimately entire pasture ecosystems. The most well known and widely documented example of this particular form of re-wilding is the Oostvaardersplassen experiment in the Netherlands (Vera, 2009a, see also Lorimer and Driessen, 2011). Formerly back-bred 'Heck' cattle and Konik horses have been introduced into a restored area of polder grassland as proxies partly in order to demonstrate how historic wild ungulates, in combination with other herbivore and bird species, could have maintained a natural grass-based northern European landscape against the development of closed-canopy forest (thereby showing that such a landscape is not the result of agricultural techniques), but partly also to understand the evolving population dynamics of such a newly re-wilded species/landscape assemblage (Vera, 2009b). Other examples of re-wilding through de-domestication might include the Chillingham herd (Hall, 1989) as well as the growing number of heavily mediated stories of former zoo, pet and circus animals being 'returned' to their 'natural' habitats, for example, the whale who 'starred' in the *Free Willy* films (Brydon, 2006) or the infamous 'lion cub from Harrods' (Bourke and Rendell, 2010).

The transgressive biopolitics of rewilding

Re-wilding and restoration are problematic. They raise considerable conceptual, ethical, governance and security issues. Commenting on their study of the Heck cattle, Lorimer and Driessen argue: 'The fluid biopolitics of re-wilding encounters fixities and frictions when it runs into other modes of [bovine] biopolitic' (2011, p. 2). At one level, to re-wild is to un-wild or even de-wild. Maskit (1999) argues that the wild is a place, not an abstraction and, for a place to be defined, it must be known. In that very act of knowing is a negation of the truly wild. If nature and culture are to be considered as 'fundamental categories of thought', then 'restoration is either invisible or repellent because it violates these basic categories', the restored landscape being a 'landscape of ambiguity' (Jordan, 2000, p. 24). Ambiguity, artifactuality and monstrosity: Lorimer and Driessen, again, revel in the 'monstrous promise of re-wilding' (2011, p. 2). The impure re-wilded and restored transgress the 'fundamental categories' of Jordan and others. In doing so, they reveal what Whatmore and Thorne refer to as the 'more promiscuous topologies of wildlife' (1998, p. 437).

As one might expect, there has been much debate over the conceptual and ethical foundations – and challenges – of re-wilding and restoration (see, for example, Callicott and Nelson, 1998; Gobster and Hull, 2000; Nelson and Callicott, 2008; Lowenthal, 2009). Eric Katz (1992; 1996) and Robert Elliot (1997) have long been outspoken critics, arguing that the human intentionality of re-wilding and restoration negates 'natural value', creating fake, domesticated or artefactual simulacra of 'Nature'. Re-wilding is both spatially and temporally anachronistic and mimetic. It locates itself within an ontological confidence in the distinctiveness of 'Nature' yet, by its very practice, encultures 'Nature' at every turn. For all its genuflection to the purity of 'wilderness', it is, finds Birch (1998), a celebration of anthropocentrism and human imperialism: 'agriculture in reverse', as Jordan (2000) calls restoration, is still culture and its products, curious hybrids and erstwhile 'monsters' (Lorimer and Dreissen, 2011). And yet, re-wilding is vigorously pursued, at a whole variety of scales. The IUCN's 1998 *Guidelines for Reintroductions*, reflecting the increase in re-wilding and reintroduction programmes, were drafted:

> in response to the increasing occurrence of re-introduction projects worldwide, and consequently, to the growing need for specific policy guidelines to help ensure that the re-introductions achieve their intended conservation benefit, and do not cause adverse side-effects of greater impact.
>
> (IUCN, 1998, p. 5)

This is not the time to fully consider the ethical and conceptual issues of re-wilding and the management of the 'neo'-wild. My interest here is with the potential and the very real impact of those 'adverse side effects' upon the latest bio-political paradigm, that of biosecurity. For, despite its rhetoric of exuberance and hopeful natural fecundity, re-wilding is really all about boundaries, both their presence and their transgression. Although Birch (1998) might define wilderness reservations – the spaces of re-wilding – as 'holes and cracks, as "free spaces" or "liberated zones" in the fabric of domination and self-deception that fuels and shapes our main-stream contemporary culture' (p. 466), there is no doubt that re-wilding and species translocation brings or has the potential to bring, to use Mike Davis's (2005) phrase, the monsters closer to our door.

Re-wilding and biosecurity

Species reintroduced as part of ecological re-wilding programmes (as distinct from those introduced for purely ornamental or hedonistic reasons) are not necessarily invasive species but they can be when they or their fellow travellers 'get out' or 'get in'. Here, one might argue, biosecurity and biodiversity (in the form of re-wilding and species reintroduction) share, what Watts (2000) refers to as the dominant paradigm of modern human–animal relations, that of physical enclosure. In reality, of course, it has proved very difficult to 'enclose' nature and natural processes whether it be for biosecurity or for biodiversity.

The recent history of species reintroduction (sometimes specifically for reasons of bio-control) and of re-wilding is littered with well known and less well known examples of translocated and imported wildlife and their associated bio-accomplices (from zoonotic pathogens to ecological stable-mates) impacting negatively upon domesticated species and indigenous wild species and landscapes, as well as human social and economic activities and practices through their displacement.

Possibly the most well known is the cane toad (*Bufo marinus*), described as 'the biggest ecological disaster in Australia' (Franklin, 2006, p. 160). Deliberately introduced from Hawaii in the 1930s as a bio-control measure against the cane beetle (*Dermolepida albohirtum*), itself a significant threat to the Australian sugar industry, the cane toad has spread across Australia in vast numbers. Poisonous to many would-be predators, such as frog-eating snakes, and an ecological competitor to other amphibian and reptile species, the toad is having a measurably dramatic effect upon local ecologies and indigenous Australian wildlife (Taylor and Edwards, 2005). Although their impact upon human social and economic activities is less clear, they are seen as a major threat to the ecological biosecurity of Australia and surrounding lands, including New Zealand.

The reintroduction of wolves to France's Mercantour National Park has been particularly controversial, one reason being that this is also an area of extensive sheep farming and recreational hunting (Mauz, 2005; Buller, 2008). Understandably, the wolves, which number around 60 individuals, have found the sheep flocks, extensively pastured on the mountain grasslands, an easy and accessible food source. A system of compensation has been established to reimburse flock owners for killed sheep and lambs, but there is growing concern that extensive sheep farming and wolves are incompatible within the national park. Experimental barriers and fences, restricting the wolves to specific areas, have not proved effective. Fencing in the sheep, overnight, has become more common but flies in the face of the traditional practice of extensive grazing. Significantly perhaps, the most effective biosecurity measure against the wolves has been the introduction of various breeds of guard dog including the Pyrenean patou, the Tibetan mastiff and the Turkish sheepdog. One reintroduction, from northern Italy to southwest France, has generated a number of others.

For the most part, however, it is the ecological configuration of the host territory that acts as a natural boundary for reintroduced species in re-wilded areas. In their guidelines, the IUCN state that such territories 'should be within the historic range of the species' (1998, p. 7). Determining that 'historic range' is not always easy. Not only does it fail to account for the adaptability of species themselves, but it also fails to account for ecosystem changes beyond the 'historic range'. The reintroduction of Norwegian beavers (*Castor fiber*) to Scotland in 2009 is a case in point, raising concerns for the impact of the beavers on wildlife in a range of Scottish rivers and beyond the woods of Knapdale, their reintroduction site (Macdonald et al., 1995).

Wild species escaping from reintroduction sites, ravaging local ecosystems, destroying farmland and attacking humans might constitute part of the alarmist narrative critical of re-wilding, but is rarely borne out in fact. Potentially of greater concern is the association of wild and re-wilded places with disease, infection and pathogenic contamination. In his account of the emerging avian flu pandemic of 2003–2004, Mike Davis (2005) makes that association explicit: 'The most ferocious of man-eaters is an innocuous companion of wild ducks and other waterfowl', he writes (p. 9) and, elsewhere, in a chapter entitled 'Birds of Hong Kong', he establishes the link between the wild birds of the 'internationally important' protected Mai Po marshes in Hong Kong harbour and the first instances of the disease.

However, while there are clearly established pathways of disease spread between wild areas and translocated species on the one hand and domestic species, indigenous wildlife and human beings on the other (e.g. Woodford and Rossiter, 1993), the potential of re-wilding projects to threaten biosecurity through pathogenic transmission has not been a major area of debate (Cunningham, 1996), though the much vaunted project of Pleistocene re-wilding in the USA has prompted concern that altered disease ecology (Dazak et al., 2000; Donlan et al., 2005b) will have potentially major human health implications. At another scale, Lorimer and Dreissen (2011) point to local farmers' concerns that the carcasses of dead Heck cattle, intentionally left to decompose in situ, will become a source of disease for their own domesticated herds.

For many, of equal importance are the biosecurity and welfare of the introduced or conserved species (Deem et al., 2011). The reintroduction of Norwegian beavers to Scotland, mentioned above, was marred by the death of five of the 17 animals during the mandatory quarantine period. Many wildlife reintroduction programmes have recorded high mortality rates amongst the reintroduced individuals (Scott et al., 1999). Reintroduced big cats in Florida wildlife areas have contracted diseases caught from domestic species. In his account of his own role in the management of the wolf population of Isle Royale, Peterson (1995) reflects on his decision not to vaccinate members of the diminishing wolf population against the disease that was killing them, despite the ecological value of doing so (see also Jamieson, 2008). Then there is the welfare of the wildlife into whose midst translocated species are placed. 'Is it acceptable', asks Bekoff (2006, p. 226) 'to do a project in which a non prey species (e.g. coyotes in Yellowstone) will be killed by the reintroduction of a competitor (e.g. grey wolves)?'

Finally, we might also acknowledge that re-wilding and reintroduction programmes can offer biosecurity solutions. The decline of the US wolf population, for example, has been linked to the rise in Lyme disease in North America (Estes, 2002). Their reintroduction thus takes on a function of human health management. There are, therefore, many openings for complicity rather than an uncritical opposition, between biodiversity (in the form of re-wilding) and biosecurity (as restriction and control). Of pathogens, Clark (2007) suggests:

> for all their terrible toll, we might also acknowledge a kind of 'generosity' in the way that pathogens take advantage of the proximity and porosity of larger bodies.
>
> (p. 65)

Re-wilding, science and environmental governance

In this final section, I want to briefly consider the tensions between re-wilding and biosecurity as forms of environmental governance. At one level, there are clear parallels between the two. Both are normative and interventionist, each in its own way, an ordering and a bio-technological strategy over 'Nature' for what are predominantly humanistic ends. These ends might be, as Jamieson (2008) would have it, nostalgic or teleological, but they are also economic. 'Biological and evolutionary stories', Haraway (2000, p. 55) reminds us, 'are so thickly layered with the tools of political economy'.

Second, both re-wilding (including reintroduction) and biosecurity are expressions of bio-control involving often similar procedures for monitoring and observation, classification, ordering, confinement and restraint (Youatt, 2008; Collard, 2012). Thus barriers, cameras, satellite tracking, samples, cages, culling, quarantine and vaccination, as well as the politics and apparatus of risk management, form a common lexicon that places these different constituents of life under increasingly global levels of surveillance. Both share a concern to 'speed up' natural processes within the framework of human generational time (Dobson et al., 1997), thereby robbing ecologies and creatures of their more 'natural' and independent histories. Here too, biosecurity and re-wilding collectively represent arenas in which human–non-human relations are becoming increasingly scientized and politicized not only through the technologies and procedures of intervention but also through forms of governance, institutionalization, authority and control (Davis and Slobodkin, 2004; Light, 2000).

Third, and perhaps less obviously, we might draw on Shukin's (2009) notion of 'biomobility' which she defines, quoting from Elder et al. (1998, p. 81), as:

> A condition in which by virtue of the 'radically changing time-space relations that epitomize postmodernity', interspecies exchanges that were once local or 'place-specific' are experienced as global in their potential effects.
>
> (Shukin, 2009, p. 183)

Re-wilding, as we have seen, frequently involves transcontinental relocation, increasingly managed by globally articulate organizations supported by global flows of scientific knowledge and finance (Whatmore and Thorne, 2000).

However, while we might draw attention to the translocation of reintroduced species in re-wilding programmes, we might fail to notice that many of the domesticated farm animals – more or less protected by biosecurity measures in increasingly specialized mono-species (and in many cases mono-gendered) spaces – are themselves far from any 'historic range' they might have had, if indeed they ever had one at all. Translocated 'wild' animals, the subjects of reintroduction, will have experienced degrees of human contact and intervention that are barely compatible with the epithet 'wild', but are far more common to that experienced by domesticated food or pet animals. Both 'biosecurity' and 'biodiversity', in these contexts, are not without their own inherent semiotic contradictions.

Nevertheless, the agendas and practices of biosecurity and re-wilding differ too in important ways: the first lies in the possibility of a politics of alterity; the second in the differential politics of life and 'life itself' (Franklin et al., 2000). Re-wilding can be controversial. The Chicago Restoration controversy stands as an exemplar of the often contested dimensions of landscape and ecological re-wilding, while various large carnivore reintroduction programmes, from Yellowstone to the Mercantour in France, have engendered powerful protests from established interest groups. In Chicago, though their motives were varied, those opposed to the re-wilding of naturally reconstituted but non-native woodland with artificially reconstituted native prairie argued that:

> the natural beauty of the unmanaged forest was being replaced by a beauty that was more manipulated and manicured, like one would find in a garden.
>
> (Gobster, 1997, p. 35)

Here though, the proponents of restoration comprised a more radical constituency drawn not only from science but also from 'alternative' politics (Stevens, 1995) articulating 'traditional' knowledge (Raish, 2000), even if this was not always so perceived.

> How did what a few thousand volunteers in the Chicago area saw as a positive, nonconfrontational, grassroots movement find itself painted as a conspiracy involving big government and allegedly secretive environmental organizations?
>
> (Siewens, 1998, p. 9)

In the French Alps, the farmers and hunters opposed to the wolf's reintroduction similarly referred to the 'secure' ecology and natural beauty of the sheep-pastured meadows and species-rich biodiversity, some of which was the direct result of earlier ungulate reintroduction programmes (Buller, 2008). The Environment Ministry and the Mercantour National Park authority, however, initially drew heavy criticism from both within the French political administration and from opposing interest groups for what was seen as their clandestine support and undemocratic facilitation of reintroduction as well as their closeness to pro-reintroduction environmental organizations (Assemblée Nationale, 2003).

What the deep ecology thought behind much US re-wilding rhetoric (Foreman, 2004), the nostalgic natural historicism of English otter reintroduction (Darlington, 2012) and the militant environmentalism of the French wolf return (Campion-Vincent et al., 2002) have in common is a celebratory commitment to a revitalized naturalism. The re-emergence of wild species is seen as generative, dynamic and ultimately boundless. In France, the Mercantour wolves have become symbolic of a new metropolitan engagement with the natural 'bio' – one that draws comfort from the fact that such wild spaces, with their charismatic wild species, are still here in a world where human–nature relations are otherwise dominated and overwritten by the hubris of security, control and management. For Haraway (2000, p. 92) 'life' is 'developmental, organistic temporality'. Of the wolves, Mauz writes: 'certitudes waiver in the face of the general production of incertitude … mastered know-how gives way to improvisation' (2006, p. 161), while Hintz argues that 'Life would be much richer, much wilder, if we worked to grant nonhuman nature more autonomy, to foster the free-flow of ecological and evolutionary processes' (Hintz, 2007, p. 186).

References

Assemblée Nationale (2003) *Rapport au nom de la commission d'enquête sur les conditions de la presence du loup en France et l'exercice du pastoralisme dans les zones de montagne*, vol. 1. Report of the Commission of the National Assembly, number 825, 2 May, Paris.

Bekoff, M. (2006) *Animal Passions and Beastly Virtues*. Temple University Press, Philadelphia, PA.

BES (2007) *The Ecological Consequences of Wilding as a Long Term Conservation Strategy*. Report of a meeting of the British Ecological Society Conservation Ecology Special Interest Group held at Gregynog, Powys, UK, 12–13 July. BES, London.

——— (2009) *Response Regarding Request for Position on Re-wilding*. British Ecological Society, London.

Birch, T. (1998) The incarceration of wilderness: the wilderness as prison. In B. Callicott and M. P. Nelson (eds) *The Great New Wilderness debate*, University of Georgia Press, Athens, pp. 443–470.

Bourke, A. and Rendell, J. (2010) *A Lion Called Christian*. Bantam Books, London.

Bowman, D. (2012) Conservation: bring elephants to Australia. *Nature*, vol. 482, no. 7383, p. 30.

Bradshaw, A. (2002) Introduction and philosophy. In M. R. Perrow and A. J. Davy (eds) *Handbook of Ecological Restoration, Volume 1: Principles of Restoration*, Cambridge University Press, Cambridge, pp. 3–9.

Brydon, A. (2006) The predicament of nature: Keiko the whale and the cultural politics of whaling in Iceland. *Anthropological Quarterly*, vol. 79, no. 2, pp. 225–260.

Buller, H. (2008) Safe from the wolf: biosecurity, biodiversity, and competing philosophies of nature. *Environment and Planning A*, vol. 40, pp. 1583–1597.

Bunce, M. (1994) *The Countryside Ideal*. Routledge, London.

Cairns, J. (2002) Rationale for restoration. In M. R. Perrow and A. J. Davy (eds) *Handbook of Ecological Restoration, Volume 1: Principles of Restoration*, Cambridge University Press, Cambridge, pp. 10–22.

Cairns, J. and Heckman, J (1996) Restoration ecology: the state of an emerging field. *Annual Review of Energy and the Environment*, vol. 21, pp. 167–189.

Callicott, J. B. and Nelson, M. (1998) *The Great Wilderness Debate*. University of Georgia Press, Athens.

Campion-Vincent, V., Duclos, J. C. and Abry, C. (eds) (2002) *Le fait du loup. De la peur à la passion: le renversement d'une image*. Special issue of *Monde alpin et rhodanien*. Centre Alpin et Rhodanien d'Ethnologie, Grenoble.

Choi, Y. D. (2007) Restoration ecology to the future: a call for new paradigm. *Restoration Ecology*, vol. 15, no.2, pp. 351–353.

Clark, N. (2007) Animal interface: the generosity of domestication. In R. Cassidy and M. Mullin (eds) *Where the Wild Things Are Now: Domestication Reconsidered*. Wenner-Gren International Symposium Series. Berg Publishers, Oxford, pp. 49–70.

Collard, R.-C. (2012) Cougar–human entanglements and the biopolitical un/making of safe space. *Environment and Planning D: Society and Space*, vol. 30, no. 1, pp. 23–42.

Cunningham, A. A. (1996) Disease risks of wildlife translocations. *Conservation Biology*, vol. 10, 349–353.

Darlington, M. (2012) *Otter Country*. Granta Books, Cambridge.

Davis, M. (1998) *The Ecology of Fear*. Vintage Books, New York.

—— (2005) *The Monster at the Door*. Owl Books, New York.

Davis, M. A. (2000) 'Restoration': a misnomer? *Science*, vol. 287, p. 1203.

Davis, M. A. and Slobodkin, L. B. (2004) The science and values of restoration ecology. *Restoration Ecology*, vol. 12, pp. 1–3.

Dazak, P., Cunningham, A. and Hyatt, A (2000) Emerging infectious diseases of wildlife – threats to biodiversity and human health. *Science*, vol. 287, pp. 443–449.

Deem, S., Karesh, W. and Weisman, W. (2011) Putting theory into practice: wildlife health and conservation. *Conservation Biology*, vol. 15, no. 5, pp. 1224–1233.

Dobson, A. P., Bradshaw, A. D. and Baker, A. J. (1997) Hopes for the future: restoration ecology and conservation biology. *Science*, vol. 277, pp. 515–522.

Donlan, C. J., Berger, J., Bock, C. E., Bock, J. H., Burney, D. A. et al. (2005a) Rewilding North America. *Nature*, vol. 436, pp. 913–914.

—— (2005b) Pleistocene rewilding: an optimistic agenda for twenty-first century conservation. *The American Naturalist*, vol. 168, no. 5, pp. 660–681.

Ehrenfeld, J. and Toth, L. (1997) Restoration ecology and the ecosystem perspective. *Restoration Ecology*, vol. 5, no. 4, pp. 307–317.

Elder, G., Wolch, J. and Emel, J. (1998) Le pratique sauvage: race, place and the human-animal divide. In J. Wolch and J. Emel (eds) *Animal Geographies*, Verso, New York, pp. 72–90.

Elliot, R. (1997) *Faking Nature: The Ethics of Environmental Restoration*. Routledge, London.

Estes, J. A. (2002) Then and now. In R. L. Knight and S. Riedl (eds) *Aldo Leopold and the Ecological Conscience*, Oxford University Press, New York, pp. 60–71.

Foreman, D. (2004) *Rewilding North America*. Island Press, Washington, DC.

Franklin, A. (2006) *Animal Nation*. USNW Press, Sydney.

Franklin, S., Lury, C. and Stacy, J. (2000) *Global Nature, Global Culture*. Sage, London.

Gamborg, C., Gremmen, B., Christainsen, S. and Sandoe, P. (2010) De-domestication: ethics and the intersection of landscape restoration and animal welfare. *Environmental Values*, vol. 19, no. 1, pp. 57–78.

Gobster, P. (1997) The Chicago wilderness and its critics: III, the other side. *Restoration and Management*, vol. 15, no. 1, pp. 32–37.

Gobster, P. and Hull, B. (eds) (2000) *Restoring Nature*. Island Press, Washington, DC.

Hall, M. (2000) Comparing damages: American and Italian concepts of restoration. In M. Agnoletti and S. Anderson (eds) *Methods and Approaches in Forest History*. IUFRO Research Series 3. CAB International Publishing, Cambridge, pp. 165–172.

—— (2009) Tempo and mode in restoration. In M. Hall (ed.) *Restoration and History: The Search for a Usable Environmental Past*, Routledge, New York, London, pp. 1–9.

Hall, S. J. G. (1989). The white herd of Chillingham. *Journal of the Royal Agricultural Society of England*, vol. 150, pp. 112–119.

Haraway, D. (2000) *How Like a Leaf*. Routledge, London.

Hintz, J. (2007) Some political problems for rewilding nature. *Ethics, Place and Environment*, vol. 10. no. 2, pp. 177–216.

Hodder, K. H. and Bullock, J. M. (2009) Really wild? Naturalistic grazing in modern landscapes. *British Wildlife*, vol. 20, no. 5, pp. 37–43.

IUCN (1998) *Guidelines for Reintroductions*. IUCN Reintroduction Specialist Group, Abu-Dhabi.

Jamieson, D. (2008) The rights of animals and the demands of nature. *Environmental Values*, vol. 17, pp. 181–199.

Jordan, W. R. (2000) Restoration, community and wilderness. In P. Gobster and B. Hull (eds) *Restoring Nature*, Island Press, Washington, DC, pp. 21–35.

Katz, E. (1992) The big lie: human restoration of nature. *Research in Philosophy and Technology*, vol. 12, pp. 231–241.

—— (1996) The problem of ecological restoration. *Environmental Ethics*, vol. 18, pp. 222–224.

Lemke, T. (2011) *Biopolitics: An Introduction*. New York University Press, New York.

Light, A. (2000) Ecological restoration and the culture of nature: a pragmatic perspective. In P. Gobster and B. Hull (eds) *Restoring Nature*, Island Press, Washington, DC, pp. 49–70.

Locke, H. and Mackey, B. (2009) The nature of climate change: reunite international climate change mitigation efforts with biodiversity conservation and wilderness protection. *International Journal of Wilderness*, vol. 15, no. 2, pp. 7–13.

Lorimer, J. and Driessen, C. (2011) Bovine biopolitics and the promise of monsters in the rewilding of Heck cattle. *Geoforum*. DOI: 10/1016/j.2011.09.002.

Lowenthal, D. (2009) Reflections on Humpty Dumpty ecology. In M. Hall (ed.) *Restoration and History: The Search for a Usable Environmental Past*, Routledge, New York, London, pp. 13–33.

Macdonald, D. W., Tattersall, F. H., Brown, E. D. and Balharry, D. (1995) Reintroducing the European beaver to Britain: nostalgic meddling or restoring biodiversity? *Mammal Review*, vol. 25, no. 4, pp. 161–200.

Maskit, J. (1999) Something wild? Deleuze and Guattari and the impossibility of wilderness. In A. Light and J. Smith (eds) *Philosophies of Place: Philosophy and Geography III*, Rowman and Littlefield, Lanham, MD, pp. 265–284.

Mauz, I, (2005) *Gens, Cornes et Crocs*. Cemagref, Paris.

Mendelson, J., Aultz, S. P and Dolan, J. (1992) Carving up the woods: savannah restoration in Northeastern Illinois. *Ecological Restoration*, vol. 1, pp. 127–131.

Natural England (2010) *Otter Survey 2010*. Natural England, Sheffield.

Nelson, M. P. and Callicott, J. B. (2008) *The Wilderness Debate Rages On: Continuing the Great New Wilderness Debate*. University of Georgia Press, Athens.

New Zealand Biodiversity (undated) Biosecurity. Biodiversity Information Online. www.biodiversity.govt.nz/land/nzbs/biosecurity/index.html.

Padel, R. (2005) *Tigers in Red Weather*. Abacus, London.

Peterson, R. O. (1995) *The Wolves of Isle Royale: The Broken Balance*. Willow Creek Press, Minocqua, WI.

Raish, C. (2000) Lessons for resoration in the 'traditions' of stewardship: sustainable land management in Northern New Mexico. In P. Gobster and B. Hull (eds) *Restoring Nature*, Island Press, Washington, DC, pp. 281–298.

Robinson, Michael J. (2011) *Petition for Rule Making: Reintroduction of the Endangered Florida Panther*. Center for Biological Diversity, Pinos Altos, NM, 42pp.

Ross, S. D. (2004) Biodiversity, exuberance and abundance: cherishing the body of the Earth. In B. Foltz and R. Frodeman (eds) *Essays in Environmental Philosophy*, Indiana University Press, Bloomington, pp. 245–259.

Scott, J. M., Murray, D. and Griffith, B. (1999) Lynx reintroduction. *Science*, vol. 286, p. 49.

Shukin, N. (2009) *Animal Capital: Rendering Life in Biopolitical Times*. University of Minnesota Press, Minneapolis.

Siewens, (1998) Making the quantum-culture leap: reflections on the Chicago controversy. *Restoration and Management Notes*, vol. 16, no. 1, pp. 9–15.

Soorae, P. S. (2011) *Global Reintroduction Perspectives 2011*. IUCN, Gland, Switzerland.

Soulé, M. and Noss, R. (1998) Rewilding and biodiversity. *Wild Earth*, Fall, pp. 1–11.

Stevens, W. K. (1995) *Miracle Under the Oaks: The Revival of Nature in America*. Pocket Books, New York.

Sutherland, W. J., Adams, W. M., Aronson, R. B., Aveling, R., Blackburn, T. M. et al. (2009) One hundred questions of importance to the conservation of global biological diversity. *Conservation Biology*, vol. 23, no. 3, pp. 557–567.

Taylor, R. and Edwards, G. (2005) *A Review of the Impact and Control of Cane Toads in Australia*. A Report to the Vertebrate Pests Committee from the National Cane Toad Taskforce. Canberra, ACT.

Vera, F. (2009a) The shifting baseline syndrome in restoration ecology. In M. Hall (ed.) *Restoration and History: The Search for a Usable Environmental Past*, Routledge, New York, London, pp. 98–110.

—— (2009b) Large-scale nature development – the Oostvaardersplassen. *British Wildlife*, vol. 20, no. 5, pp. 28–36.

Watts, M. (2000) Afterword: enclosure. In C. Philo and C. Wilbert (eds) *Animals Spaces, Beastly Places*, Routledge, London, pp. 292–303.

Whatmore, S. and Thorne, L. (1998) Wild(er)ness: reconfiguring geographies of wildlife. *Transactions of the Institute of British Geographers*, vol. 23, no. 4, pp. 435–454.

—— (2000) Elephants on the move: spatial formations of wildlife exchange. *Environment and Planning D: Society and Space*, vol. 18, pp. 185–203.

Wilson, A. (1992) *The Culture of Nature: North American Landscape from Disney to the Exxon Valdez*. Between the Lines, Toronto.

Woodford, M. H. and Rossiter, P. B. (1993) Disease risks associated with wildlife translocation projects. *Revue Scientifique et Technique (International Office of Epizootics)*, vol. 12, no. 1, pp. 115–135.

Youatt, R. (2008) Counting species: biopower and the global biodiversity census. *Environmental Values*, vol. 17, no. 13, pp. 393–417.

13

THE INSECURITY OF BIOSECURITY

Remaking emerging infectious diseases

Steve Hinchliffe

Introduction

A few decades ago there was widespread promise of a third epidemiological transition, signalling a global decline in contagious diseases (Omran, 1971). And yet around 200 new infectious diseases were identified in the last two decades of the twentieth century (Taylor et al., 2001). The most significant by far of these emergent conditions was AIDS (acquired immuno-deficiency syndrome), causing 1.8 million deaths in 2010 (UNAIDS, 2011). Much less significant in terms of numbers of deaths were diseases like SARS (severe acute respiratory syndrome), which quickly moved from Asia to North America in 2003 and caused something like 1,000 deaths (WHO, 2003). Like avian influenza and swine flu, this disease became famous less for the number of associated human deaths and more for the speed with which it moved around the planet. But it was the potential of these diseases to become even more virulent and dangerous that caught the eye. For many commentators they ushered in a new pandemic age. Even sober analysts suggested that the 'world is teetering on the edge of a pandemic that could kill a large fraction of the human population' (Webster and Walker, 2003, p. 122).

An overwhelming feature of the suite of newly emerging or re-emerging infectious diseases (EIDs) is that most (c. 75 per cent of 200 or so that have been identified) are zoonotic, that is they have crossed from nonhuman animals to people and possibly back again. This relationship between people and animals, or this sharing of disease conditions, is not new. All influenza viruses for example have avian components. Perhaps we shouldn't be surprised at these prolific crossings from nonhumans to humans. Biologically, humans are hardly exceptional, and share a good deal of their genetic and cellular structures with the vast majority of life on earth. Moreover, the evolutionary role of microbes in shaping human beings is becoming more and more apparent (Margulis and Sagan, 1986). As geographers and others have repeatedly demonstrated, the world is 'more than human' and, as a corollary, life and disease are always more than biological. They are characterized by intense couplings of human and nonhuman, bodies and technologies, animal and machine, such that crossings between species may be the norm rather than the exception (Haraway, 1991; Latour, 1993; Whatmore, 2002; Hinchliffe, 2007; Bennett, 2010).

But it is still pertinent to ask the question, why at this juncture should we fear emerging diseases? This chapter provides some answers, looking at zoonotic diseases in terms of emergence, infection and transmission processes. Each of these is always more than biological and, to emphasize this point, emergence, infection and transmission are mapped onto ecologies of production, social ecologies of resilience and circulations, respectively. This mapping allows for a description and subsequent evaluation of government- and industry-sponsored responses to emerging zoonotic diseases. The latter are often bundled together under the headings agricultural modernization and biosecurity, and they tend to involve the *integration* of so-called secure practices and their subsequent *separation* from insecure spaces. The focus is on poultry in the UK, and research material is drawn from fieldwork on farms, in processing plant, in laboratories and across the food system.[1] The argument is that biosecurity may actually increase the potential for catastrophic emerging infections. I argue that a paradox is at play whereby the very process of making life safe or secure can generate new insecurities.

Emerging infectious diseases: why now?

Emergence suggests a relentless process of co-evolution, producing new forms as a result of a continuous mixing of organisms, hosts and environments (Cooper, 2006). This intermixing can generate new viral or bacterial strains, but it is often alterations in the relations of hosts, environments and microbes that conspire to generate new conditions of possibility for disease. The virologist Stephen Morse demonstrated, for example, that many of the newly emerging diseases at the tail end of the twentieth century were not necessarily related to new viruses *per se*, but were more often the result of new crossings from nonhuman animals to people. A main reason for this emergence and re-emergence of infectious diseases was the recent enlargement of what he called the zoonotic pool, the available set of possible diseases that could cross from nonhuman to human populations (Morse, 1993). Changes to the pool were, he argued, largely the result of the land use changes being wrought by humans across the planet. The argument was that, as wildlife is disturbed through deforestation and urban and agricultural expansion, then the chances for human–animal interaction are increased. This large-scale displacement produces new potential interfaces, proximities and intensities, changing the relations between animal and human bodies.[2]

Alongside displacement, a key component of this expanding zoonotic pool is livestock. Domesticated and semi-domesticated animals are often in direct and indirect contact with wild animals, acting as the first recipients and then translators of wild-type micro-organisms. These translations result from 'routine' mutations of micro-organisms, which may generate the right protein conformations that make infection of people more successful and subsequently evolve into new and possibly dangerous strains. Perhaps more alarmingly, co-infections of livestock with different bacterial and viral types can lead to reassortments, or effectively a mixing up of genetic information, with the result that new and possibly highly pathogenic micro-organisms can be produced almost immediately. This potential for the rapid emergence of newly infectious diseases is aided not only by increased wild–domestic animal and human interactions, but also, some argue, by the sheer mass of livestock animals that now form a key component in the resourcing of an expanding and protein-demanding human population.

Indeed, for some commentators, it is the changes in farming practices over the last few decades, the rise of a global meat protein market and the concomitant rise in global animal production that has inflated the zoonotic pool, or at least, in the colourful terminology of Mike Davis, provided the engine room for or amplifier of a global zoonotic disaster (Davis, 2005; Liebler et al., 2009). Certainly, if we take poultry as an example, the estimated 52 billion birds that are slaughtered per year, worldwide, gives a sense of the standing population of microbial amplifiers for viral respiratory diseases like avian influenza or bacterial food-borne diseases associated with *Campylobacter*. Add to this the overwhelming tendency for most of these chickens to be almost genetically identical and housed together in close proximity, and the notion of a zoonotic pool becomes more vivid. Moreover, the pool is expanding. The 2008 collapse of global property markets and the simultaneous rise in worldwide food and commodity prices contributed to an investment spike in food, and particularly animal protein, production. Finance capital was diverted to concentrated animal feeding operations (CAFOs) in newly emerging protein production zones, particularly in Asia (Wallace, 2009).

Disease emergence, even in this brief treatment, links together financial institutions, consumer habits, population growth, environmental change, large food corporations, viruses, genetic and breeding technologies, chickens, and so on. Even before we discuss infections and transmissions, it should be clear that emergence is more than a biological phenomenon. It relates to what Wallace (2009) has called a political virology and it underlines a critical role for the social sciences within biosecurity debates and research (see also Scoones, 2010; Scoones and Forster, 2011). Any intervention in making life safe, or biosecurity, requires that we not only attend to the interactions of wild and domestic animals, and people and animals, but also consider the *ecologies of production* that actively play a role in disease emergence.

Within this political virology, it is important to understand not only the potential of viruses and bacteria to criss-cross species boundaries. It is also important to understand how the effect of those micro-organisms is dependent on far more than their properties alone. This brings us to the second term in the list, infection. Disease is always more than a matter of infection: it is a pathogenic entanglement of hosts, environments and microbes, a relational achievement (Hinchliffe, 2007). Viruses and other micro-organisms can only infect or certainly can only cause illness in susceptible and vulnerable bodies. While classical germ theory invites us to ascribe the causes of disease to the active microbial agent, this ascription is always a partial one that is achieved by foregrounding the microbe while relegating the host and environment to the background.[3] Once disease is treated as a relational achievement involving microbes, bodies and environments, it follows that infections will be a function of disease vulnerability as well as pathogens. Patterns of health and disease are therefore directed as much by the social life of, or associations between, people and animals as they are shaped by biology. So, in addition to the ecologies of production, we also need to consider the *social ecologies of resilience*.

The term resilience is derived from ecology and broadly means the relative ability of a system to cope with perturbations or changes.[4] In a broader sense though, it can be understood as the relative vulnerability of a social and ecological organization. Reductions in resilience to emerging infections might be treated as roughly synonymous with declines in the ability of socio-ecologies to deal with new challenges. Rises in global poverty will, for example, produce a breeding ground for disease. Rapid changes to ecosystems, climate change and reductions in biodiversity can also reduce resilience. Less obviously, perhaps, an increasingly interconnected and tightly coupled food production system, where diversity is actively discouraged, presents a rich field for microbial activity and so may increase vulnerability. Resilience in this sense is more than the ability to fight off outside threats. There is also the degree to which perturbations within a system are transmitted or defused. Here, a tightly coupled food production system might have more in common than would be first imagined with a highly engineered power station, less susceptible to external threats but more likely, as a result of feedbacks and system loops characterized by rapid flow, to quickly produce failure (for the classic account see Perrow, 1999; for an account of foot-and-mouth disease along these lines see Law, 2006).

Finally there is transmission of disease, a process that is augmented by the manifold circulations that characterize economies and social lives. Diseases have always travelled and have followed empire and trade for centuries. Arguably though, contemporary viral traffic can fold together previously distant places with a speed and volume that has not been experienced before. The ability of a subclinical carrier to board a plane in Amsterdam and inadvertently transmit an infection to Buenos Aires or Tokyo, well before clinical symptoms set in, is well known. Enhanced connectivity alerts us to the rapid movements of people, animals, plants and micro-organisms, and the ability for the movement to occur long before a host organism has developed clinical symptoms. But connectivity also, to echo the previous paragraph, alerts us to the imbrications of global systems from pharmaceuticals to finance, from world trade to the farm gate, from migrations to climate change. Moreover, it is the manifold circulations of liberal economies, both the ideal of free circulation but also the increased specialization and spatial divisions of economic activity, that start to generate new conditions of possibility for disease. It is this spatially coupled and folded global landscape that forms the context for the following evaluations of the main responses to emerging infections, biosecurity.

Biosecurity in practice

The potential for newly emerging and re-emerging infectious diseases to generate high impacts presents a key challenge for governments and other corporations. These catastrophic risks (Ewald, 1993) are characteristically met with a mixture of responses including prevention, preparedness and pre-emption. Here I will focus on prevention, saying something about preparedness and pre-emption in the closing section.

Disease prevention remains a central component of biosecurity or making life safe in light of EIDs. Prevention is a preferred medical and economic paradigm based on the mantra that an ounce of prevention is often worth a pound of cure. But prevention is necessary in complex societies because, if danger exists,

> it exists in a virtual state before being actualised in an offense, injury, or accident. This entails the further assumption that the responsible institutions are guilty if they do not detect the presence, or actuality, of a danger before it is realised.
>
> (Ewald, 1993, pp. 221–222)

In other words, prevention involves organizations anticipating a danger, both in the sense that a potential threat can be demonstrated, but also in the sense that various responsible organizations are required to perform their responsibility through clear acts of disease prevention. Arguably it is the latter that colours or shapes state and industry approaches to biosecurity. What ensues is a characteristic responsibility game whereby, in the UK scene at least, the livestock industry is increasingly organized through the wholesale and retail industry and becomes ever more conscious of the requirement to act on behalf of consumers in preventing disease from reaching the supermarket shelves. Prevention, in these circumstances, takes on the following features: increased managerial integration of production ecologies, wild/domestic interface management (or barrier systems) and disease surveillance. I will now outline what is involved in these approaches, starting with integration.

It is often assumed that better biosecurity is made possible through integration of processes such that disease-free status can be assured from farm to fork. In this case, farms are increasingly under the influence of, or have their management circumscribed by, actors further up the food chain including processors, retailers and marketing organizations. This may include outright ownership of farms by processing companies, managerial integration through contracts with retailers or wholesalers, and encouragement, under sanction of exclusion from particular markets, to comply with various accreditation schemes that specify among other things the disease status of herds or flocks. Further integration is effected through consolidation and expansion, with larger operations assumed to be able to generate biosecurity efficiencies in terms of the size of their operation, their uniform processes and the ability to absorb initial costs of compliance.

To take an example, the UK poultry industry is dominated by a handful of processing companies, each with contracts to major retailers and either managerial- or ownership-based vertical integration of the chicken production process (from farm to wholesale). Major companies supply the breeding stock, chicks and feed, and the chickens are 'grown' (as it is termed) under specified management systems. As one representative of a major company notes: 'we are very much in control of the growing programmes and the agricultural side' (Interview with poultry processor, 2011). This integration of actors is largely organized around the extension of sanitation to all parts of the food growing process. As the retailers, who increasingly drive this process, capture it, biosecurity takes 'practices we would use in a food factory where segregation is critical … and applies them into agriculture' (Retailer interview, October 2011). This is integration through the process of bio-sanitary extension, where farms become ever more materially and symbolically distinct in terms of their bio-ecologies from their surrounding environment. As such, there is a metaphorical and literal closure of the barn door, to enclose livestock and dead stock in a world of regulated microbial (in)activity.

The ensuing barrier systems include the structural integrity of buildings, their regular maintenance and general design to prevent incursion of pathogens (Dent et al., 2008). Techniques include: reduced contact between livestock and farm workers through automated feed systems; antechambers or clean areas for changing clothes, with staff to don, as one set of model farm instructions says, 'freshly laundered overalls each day'; exchange sites where removals of dead stock can be undertaken at the perimeter of the farm; and various technologies for fly and rodent control. Needless to say, transforming aged and often ad hoc farm infrastructure to conform to these security requirements is expensive and provides its own momentum for buy-outs, amalgamation and integration of production.

Finally, in order to ensure that the management of interfaces between wild and integrated domestic enclosures is effective, disease surveillance both within and outside the enclosure is undertaken. In other words, integration and separation requires continuous policing in order to provide assurances of disease-free status and early warning of any proximate threats to that status. Surveillance verifies closure, confirming 'health' (or absence of pathogens) on the inside and allowing for potential infringements or challenges from outside to be picked up quickly so that disease events can be contained. Again, then, the investment in biosecurity is predominantly focused upon disease incursion and contamination of livestock production systems and beyond this the dangers to public health. This is a walling-in of 'good' life and a walling-out of risky lives. The main recurring theme here is a process of upstream integration of retailers, wholesalers, farms and even wildlife, producing a policed enclosure whose success can be measured by an ability to limit flows of undesirables across territories and bodies. In the next section I will review that success not only in terms of official accounts, but also by measuring it against the ecologies of production (emergence), resilience (infection) and circulation (transmission) that, I have already argued, are central to the making of emerging infectious diseases.

Biosecurity in question

So how do we assess the success of biosecurity in the face of emerging infectious diseases? At first blush, things look healthy in the UK. Lessons from the 2001 foot-and-mouth disease event regarding biosecurity seem to have been learned. So, taking avian influenza for example, there is a general opinion that, give or take a number of minor and short-lived outbreaks, the UK is disease-free with respect to Highly Pathogenic Avian Influenza (HPAI). Moreover, modern integrated agri-culture, with the highest of biosecurity standards and surveillance, has delivered a secure poultry ecology. This is a standard response from public health, veterinary and industry experts. Poultry and pigs, in particular, are considered to be well-managed intensive systems where biosecurity is solved, give or take the odd issue of non-compliance.

Yet, without wanting to necessarily undermine the importance of this achievement, there are several arguments that need to be raised concerning what biosecurity in this form does and doesn't address.

Emergence – and the ecologies of production

In shoring up the divides between domestic and wildlife, and more generally between inside and outside, integrated biosecurity addresses one side of the emerging infection issue. The focus is almost solely on preventing viral incursion from wild populations or from outside the integrated biosecure community. Managing the wild/domestic, inside/outside interface takes precedence over any attempt to address the conditions of production and the potential for domestic livestock to produce disease. The assumption tends to be that closing the high-tech barn door will keep disease out, making for healthy lives inside the enclosure.

And yet, the *potential* for the emergence of disease *within* the integrated, closed and biosecure system of poultry production *is* acknowledged in the surveillance process. The poultry survey, for example, is designed to act as an early warning of possible presences of low pathogenic influenzas, which might develop into high pathogenic strains, in house. In the words of the UK's 'Notifiable Avian Disease Control Strategy for Great Britain':

> The aim of the survey is to identify the *circulation* of AI viruses in poultry (in particular, waterfowl poultry species) before they become widespread in the poultry population. As such, control measures can be taken to possibly prevent *mutation* into a HPAI virus.
>
> (DEFRA, 2012, p. 12, emphasis added)

Since its inception, the poultry survey has detected H5 and H7 viruses in British poultry, confirming that flu continues to circulate, though perhaps with lower frequency in biosecure operations. Moreover, as the Strategy also makes clear, mutation of flu viruses within domestic poultry is both a possibility and a matter for control and prevention. The latter involves measures to attempt to contain or isolate the affected population until such time as they test negative for the virus. However, in all cases in the UK to date, the virus presences identified in the survey have already 'moved on', either having died out or continued their circulation before such measures could be effective.[5] In sum, a recognition of a virtual potential for in-house microbial fermentation is addressed through an aspiration to bio-containment should viral potential be actualized. But the speed with which the virus can circulate through avian bodies means that the identification of the risk would depend on regular sampling with almost blanket coverage. The expense, to say nothing of the potential for samplers to spread disease, would be prohibitive. The broader point for now is that, other than through surveillance and containment after the fact, avian disease policy fails to address the potential for current forms of livestock farming to generate the conditions for viral mutation, in other words for the ecologies of production to add to the zoonotic pool.

Infection – or the vulnerabilities of sped-up lives

The poultry industry may well be relatively biosecure, in terms of barriers to wildlife and to outsiders more generally, but life behind those barriers has changed radically in the last half century. In that period the growth rates of chickens have doubled, their 'finished' weights have increased, while their life times and feed conversion rates (the ratio of feed input to chicken output) have both halved (Godley and Williams, 2008, 2009). There is, in effect, more chicken in less time with less feed. This shift to high and accelerated throughput has resulted from a variety of biological, technological and social changes, not least of which are: the breeding of an industry standard (the Vantress or Cornish Cross); the development of high protein diets; the application of pharmaceuticals to reduce common infectious diseases and to manage behaviour; the development of controlled windowless poultry housing, accelerating year-round growth; and the logistical and integrated organization of markets for a highly perishable end product (Godley and Williams, 2008, 2009). The result is cheap protein but also a life that seems constantly on the edge of 'safe'. One poultry vet extols the high standards of poultry keeping in the UK before describing the bare life of the birds in the sheds. Commenting on the tight margins that have driven poor units out of production, the vet notes:

> What's *left* in this country is a *very good* core poultry farming. Disease levels are therefore low, management is to a very high level.
> (Veterinary interview, February 2012, original emphasis)

206

Following straight on from this account of finance-driven efficiency, there's a more ambivalent statement:

> They're rearing racehorses, those birds *have to* grow and have to go through like a race, to be honest, and if they slow down at any time, that's it.
>
> (Veterinary interview, February 2012, original emphasis)

The racehorse metaphor is not so much a description of physique as one of productivity (and indeed other interviewees have referred to modern broilers as Sumo wrestlers). Nevertheless, the economic pressures and the strain on living tissues are palpable in this utterance, as chickens become high performing thoroughbreds and producers toil at the edge of profitability. Moreover, and later on in the same interview, some of the effects of these sped-up lives become apparent. In reference to possible changes in poultry house environments designed to improve welfare, the same vet says:

> The modern bird is very close to diarrhoea shall we say. You're putting a high nutritional value product in one end and you *can* tend to get looser droppings out the other end. You're growing a 2.5-kilo bird in 38/39 days, which used to take, even 10 years ago, would have been 5 days longer.
>
> (Veterinary interview, February 2012, original emphasis)

So while birds are managed more effectively, and known diseases controlled pharmaceutically, a rapidly grown bird is seemingly one handclap away from stress and possible ill-health (as the same vet adds, shouting or any interruption will cause the birds to defecate). The resulting immuno-suppression, new availability of niches for micro-organisms (now that most common types have been eradicated) and propensity for pathogens to move from the gut into muscle tissue are linked to a rise, within the last few decades, of *Campylobacter* in chicken (Humphrey, 2006; Cogan et al., 2007; Liebler et al., 2009), a pathogen that is now the most common cause of food poisoning in the UK. Moreover, there's a general sense that the integrated poultry industry may not in fact be terribly biosecure, a condition related less to the passage of disease from outside, but more to its amplification within. Indeed, the notion of biosecurity tends to obscure rather than highlight the compromised ecologies of resilience that may characterize integrated and intense poultry systems.

Transmission – or the entanglement and circulation of disease

While poultry in the UK is not as concentrated in terms of geographical location as is the case in the USA, Italy and the Netherlands, the level of processor- and retailer-led integration, as well as the transnational links within the poultry industry, make for a new set of proximities. Typically, each poultry farm is arranged into a series of poultry sheds with 30,000 chickens per shed (there is variety between farms, but large units with several hundred thousand birds as standards, high-welfare or free-range are the staple in the UK industry). Each of these farms is tightly linked into a production process and is supplied with one-day-old chicks from an integrated breeding and hatching company, and then grows chickens for 4–5 weeks prior to an initial catch or thin (the removal of 15–25 per cent of the generation in order to limit over-stocking) and then a further 1–2 weeks before sending the remaining birds to a processing plant often handling over a million birds per week. As this short description suggests, day-old chicks are effectively dispersed onto farms and then funnelled back for processing before being distributed to retailers.

This trafficking of bird life is on the face of it biosecure; the integrated business model provides for highly monitored and secure movements of birds and high standards in terms of stock. There is a kind of 'closed flock' from within the integrated system. But there are two observations to be made: first of all, system-level closure is not the same as farm-level closure. While biosecurity is most often imagined as a process of enclosure, the poultry industry reveals a world of manifold circulations, or continuous movements of living and not long since dead bodies (which are of course living in terms of microbial activity). These circulations are driven by various market and non-market relations, and combine to produce a relatively cheap, in terms of retail price, source of protein.

Second, the ability to regulate manifold circulations is often undermined by the very processes that make those circulations necessary. So, for example, as poultry production becomes scalable through capital investment in processing plant and CAFOs, one of the inputs that tends to be omitted from this scaling, or outsourced, is labour. While CAFOs tend to be staffed minimally, often with a single farm manager and assistant to regulate inputs and manage shed conditions on a daily basis in order to ensure good waste management (everything from ensuring the litter is dry to removing dead and diseased stock), there are points in the process when more labour is required. Key here is the poultry catch, when sub-contracted teams of catchers move from farm to farm in order to catch birds and load them into crates and large trucks prior to shipping to slaughter. The catches effectively and inevitably expose human and avian bodies to one another in conditions that are time-pressured, hot and undoubtedly disturbing for immuno-compromised birds. In standard production systems birds are caught by hand at a rate of 1,500 per hour. Each catcher picks up 6–7 chickens by their feet before transferring them to the modules. There can be little doubt, as one microbiologist and industry expert put it, that this induces stress in birds and the stress generates its own mechanisms for the proliferation and circulation of disease. So, the levels of *Campylobacter* in chicken muscle, for example, tend to rise between farm and slaughterhouse:

> That is because of the stress of catching, being put in the crate, being starved, and one of the things that happens ... their gut is flooded with

noradrenaline … [the] noradrenaline captures iron and takes it to *Campylobacter* and increases its growth rate by about tenfold. That, we think, explains the difference between transported and non-transported animals, and because the iron that the bug has now got has switched on virulence genes, this bug, if you give it another chicken, is much more invasive.

(Industry expert interview, May 2011)

The crates full of birds and their excrement are loaded onto open trailer trucks (5,500 birds per lorry). After a careful hosing of tyres with standard viricide (a somewhat symbolic ritual passed down from the 2001 foot-and-mouth epizootic), the chickens make the cross-country journey to the slaughtering and processing plant. The other bodies on the move are the catchers. Paid on a piecework basis and, at the time of writing, with little or no pay incentives to sign in sick if they are unwell, these bodies, like the people working in the processing plant, mark the frontline for circulation of microbes that are shed by stressed avian lives.

With these circulations and interfaces in mind, biosecurity cannot be imagined as a matter only of enclosure. Life, even the bare life of contemporary broilers, circulates. In addition, the more we enclose those lives the more we may, inadvertently, ensure that their circulations have even greater potential to generate new threats.

Conclusions: the insecurity of security

In this chapter I have discussed the conditions that lead to emerging infectious diseases and have in the process highlighted livestock, ecologies of production, social ecologies of resilience and circulation as key concerns. Using fieldwork in the UK, I have characterized biosecurity within the food and farming sector to be understood as a matter of greater integration and separation of farming practices from their surrounding environments. While there is some evidence that this approach has provided biosecurity gains in the poultry industry I have also argued, using field observations and interview materials, that the conditions for disease emergence, infection and transmission remain and are even encouraged within biosecure, integrated production systems. The high-throughput, circulatory lives of integrated poultry may not, in this case, be as biosecure as we would like to imagine.

The resulting need to question what is conventionally understood by biosecurity relates to the potential for security, more generally, to contribute to its own insecurities. Liberal economies are increasingly, it seems, organized through spaces of circulation (Foucault, 2007), with expansions in trade, exchange and specialization. These circulations make certain forms of life and economies flourish, but the selfsame flows can also distribute dangers. So a circulation of chickens through an integrated system not only allows for an extraction of surplus value, but can also move bacterial and viral pathogens quickly and silently. Value and plague, life and death move, as Foucault noted it, through the same circulatory systems (Foucault, 2007).

The issue then for a politics of life or biopolitics is to maintain as far as possible the life-enhancing flows while regulating those that pose threats. And as I hope this chapter has suggested, the trouble is that good and bad often share the same spaces. As a result, regulating threats can simultaneously undermine things that were beneficial. The paradox that Foucault draws out is that too much regulation, too much control, can make security counter-productive, as it interferes with the very circulations that make life flourish. His point might be summarized as: security is never finished, never solved and always potentially counter-productive. And, in this sense, modern farming, with its impossible enclosures and regulated bare lives, can add to rather than reduce the zoonotic pool. The very act of securing by enclosing can create the conditions of possibility for different kinds of events, some of which may be far more serious than would have otherwise been the case.

So, instead of asking whether or not we have full integration and compliance with a single model of biosecurity (integration and separation), we need to shift the debate to ecologies of production, resilience and circulation: to ask, in other words, what kinds of life are being enclosed and can they really contribute to safe lives, or biosecurities, in this format. Furthermore, and to touch on an issue that I have not had space to develop here, the role of conventional understandings of agricultural biosecurity in generating new and uncertain threats, or insecurities, cannot be offset by a shift to greater preparedness or pre-emptive counter-proliferations in vaccine and viral innovation. As Melinda Cooper has made clear, the new fears around infectious diseases, the political turn to emergency preparedness and to bio-tech responses, come at a particular political moment, one where the old public health models of risk and state planning for predictable episodes is being willingly and actively dismantled in favour of more speculative forms of government (Cooper, 2008). The production of an agricultural infrastructure which may be more rather than less disease prone, is occurring at the same time as a dismantling of a public disease response infrastructure. The conditions for emergence, infection and transmission may never have been so favourable.

Notes

1 The research draws on investigations into biosecurity and its interrelations with wildlife, food, labour and publics, as part of an ESRC-funded research project entitled 'Biosecurity Borderlands' (RES-062-23-1882). I am grateful to members of the project team for their input into this work and the arguments contained here.

2 The argument is now commonplace and sometimes countered by stating that, with over 50 per cent of people worldwide now urbanized, these interfaces are actually diminishing. Such an argument is made by bacteriologist Hugh Pennington (2011), an otherwise authoritative commentator on infectious diseases, but someone who clearly underestimates the biological sweep of modern cities, the animal presences in peri-urban backyard farms and in living rooms, and the daily traversing of city spaces by animals.

3 If actor network theorists have long since asked us to augment this agential ascription process by highlighting the networks that make action possible, and through this have more often than not contributed to a general decentring of human agency, here the effect of claiming microbes as relational achievements is to draw human agency and social conditions back into disease accounts. Viruses and bacteria are actors, but they are also enacted in the networks of the social; see Law and Hassard (1999).

4 Resilience has a degree of interpretive flexibility. Two variants include an engineering sense wherein the key issue is the time taken to return to a stable maximum and an ecological sense which measures the ability of an ecosystem to remain broadly coherent despite perturbations. In the latter the issue is not so much resistance to change, but the ability to live with change. See Holling (1973); and Walker and Cooper (2011).

5 The serological tests for H5 and H7 subtypes rely on the proxy of antibody presence in the blood of sampled birds, indicating a historical exposure. At the time of writing no live virus had been detected following a positive antibody result, even though the antibody traces for H5 and H7 were reasonably common.

References

Bennett, J. (2010) *Vibrant matter: A political ecology of things*, Duke University Press, Durham, NC.

Cogan, T. A., Thomas, A. O., Rees, L. E., Taylor, A. H., Jepson, M. A. et al. (2007) 'Norepinephrine increases the pathogenic potential of Campylobacter jejuni', *Gut*, vol. 56, pp. 1060–1065.

Cooper, M. (2006) 'Pre-empting emergence: The biological turn in the war on terror', *Theory, Culture and Society*, vol. 23, no. 4, pp. 113–135.

—— (2008) *Life as surplus: Biotechnology and capitalism in the neoliberal order*, University of Washington Press, Seattle, WA.

Davis, M. (2005) *The monster at our door: The global threat of avian flu*, The New Press, New York.

DEFRA (2012) 'Notifiable avian disease control strategy for Great Britain', London, www.defra.gov.uk/publications/files/pb13701-avian-disease-control-strategy.pdf.

Dent, J. E., Kao, R. R., Kixx, I. Z., Hyder, K. and Arnold, M. (2008) 'Contact structures in the poultry industry in Great Britain: Exploring transmission routes for a potential avian influenza virus epidemic', *BMC Veterinary Research*, vol. 4, no. 27, pp. 1–14.

Ewald, F. (1993) 'Two infinities of risk', in B. Massumi (ed.) *The politics of everyday fear*, Minnesota University Press, Minneapolis, pp. 221–228.

Foucault, M. (2007) *Security, territory, population: Lectures at the Collège de France 1977–78*, Palgrave Macmillan, London.

Godley, A. and Williams, B. (2008) 'The chicken, the factory farm and the supermarket: The emergence of the modern poultry industry in Britain', in W. Belasco and R. Horowit (eds) *Food chains: Provisioning, from farmyard to shopping cart*, University of Pennsylvania Press, Philadelphia, pp. 47–61.

—— (2009) 'Democratizing luxury and the contentious "invention of the technological chicken" in Britain', *Business History Review*, vol. 83, Summer, pp. 267–290.

Haraway, D. (1991) *Simians, cyborgs, and women: The reinvention of nature*, Free Association Books, London.

Hinchliffe, S. (2007) *Geographies of nature: Societies, environments, ecologies*, Sage, London.

Hinchliffe, S. and Bingham, N. (2008) 'People, animals and biosecurity in and through cities', in H. Ali and R. Keil (eds) *Networked disease*, Blackwell, Oxford.

Holling, C. S. (1973) 'Resilience and stability of ecological systems', *Annual Review of Ecology and Systematics*, vol. 4, pp. 1–23.

Humphrey, T. J. (2006) 'Are happy chickens safer chickens? Poultry welfare and disease susceptibility', *British Poultry Science*, vol. 47, pp. 379–391.

Joint United Nations Programme on HIV/AIDS (UNAIDS) (2011) 'Global HIV/AIDS response, epidemic update and health sector progress towards universal access', Joint United Nations Programme on HIV/AIDS, www.unaids.org/en/media/unaids/contentassets/documents/unaidspublication/2011/20111130_UA_Report_en.pdf, accessed 7 August 2012.

Latour, B. (1993) *We have never been modern*, Harvester Wheatsheaf, Hemel Hempstead.

Law, J. (2006) 'Disaster in agriculture: Or foot and mouth mobilities', *Environment and Planning A*, vol. 38, no. 2, pp. 227–239.

Law, J. and Hassard, J. (eds) (1999) *Actor network theory and after*, Oxford and Keele, Blackwell and Sociological Review.

Liebler, J. H., Otte, J., Roland-Holst, D., Pfeiffer, D. U., Soares Magalhaes, R. et al. (2009) 'Industrial food animal production and global health risks: Exploring the ecosystems and economics of avian influenza', *EcoHealth*, vol. 6, pp. 58–70.

Margulis, L. and Sagan, D. (1986) *Microcosmos: Four billion years of evolution from our microbial ancestors*, Summit, New York.

Morse, S. S. (1993) 'Examining the origins of emerging viruses', in S. Morse (ed.) *Emerging viruses*, Oxford University Press, Oxford, pp. 10–28.

Omran, A. R. (1971) 'The epidemiological transition: A theory of the epidemiology of population change', *The Milbank Memorial Fund Quarterly*, vol. 49, no. 4, pp. 509–538.

Pennington, H. (2011) 'The viral storm: How likely are we to die from a new pandemic?' *Guardian*, 28 October.

Perrow, C. (1999) *Normal accidents: Living with high risk technologies*, Princeton University Press, Princeton, NJ and Chichester.

Scoones, I. (2010) 'Fighting the flu: Risk, uncertainty and surveillance', in S. Dry and M. Leach (eds) *Epidemics: Science, governance and social justice*, Earthscan, London, pp. 137–164.

Scoones, I. and Forster, P. (2011) 'Unpacking the international responses to avian influenza: Science, policy and politics', in I. Scoones (ed.) *Avian influenza*, London, Earthscan, pp. 19–64.

Taylor, L. H., Latham, S. M. and Woodhouse, M. E. (2001) 'Risk factors for human disease emergence', *Philosophical Transactions of the Royal Society B: Biological Sciences*, vol. 356, pp. 983–989.

Walker, J. and Cooper, M. (2011) 'Genealogies of resilience: From systems ecology to the political economy of crisis adaptation', *Security Dialogue*, vol. 42, no. 2, pp. 143–160.

Wallace, R. G. (2009) 'Breeding influenza: The political virology of offshore farming', *Antipode*, vol. 41, no. 5, pp. 916–951.

Webster, R. G. and Walker, E. J. (2003) 'The world is teetering on the edge of a pandemic that could kill a large fraction of the human population', *American Scientist*, vol. 91, no. 2, pp. 122.

Whatmore, S. (2002). *Hybrid geographies: Natures, culture, spaces*, Sage, London.

WHO (World Health Organization) (2003) 'Summary table of SARS cases by country, 1 November 2002–7 August 2003', www.who.int/csr/sars/country/2003_08_15/en/index.html, accessed 7 August 2012.

Wolch, J. (1998) 'Zoopolis', in J. Wolch and J. Emel (eds) *Animal geographies*, Verso, London.

14

CONCLUSIONS

Biosecurity and the future – the impact of climate change

Sarah L. Taylor, Andrew Dobson and Kezia Barker

Introduction

In biosecurity rationalities and discourse, the future is brought into the realm of contemporary political calculation through risk management, as the unpredictability of life is used to justify actions made in the present to attempt to control, or produce, the future. For example, the UK's *Sunday Telegraph* newspaper reported that homeowners were refused mortgages by banks and building societies (e.g. Barclays, Santander and Lloyds Banking Group) due to the presence of Japanese knotweed (*Fallopia japonica*) in the vicinity of their homes (Gray, 2010), which can push through concrete and cause damage to buildings. Affected properties could be devalued by over £10,000 (Gray, 2010). Biosecurity approaches also respond to or produce a particular future-orientated 'affect'. This can take a variety of forms. These include the anxiety, fear and worry of producers (foresters, farmers, etc.) who wait for the next pest or disease to arrive on their land, or of people who see cherished landscapes being altered through forces apparently beyond their control (this can induce a condition for which the term 'solastalgia' has been coined – Albrecht, 2005). Then there is the excitement and passion of community groups involved in native restoration projects. For example, the Dorset Wildlife Trust held a raffle to decide which lucky volunteer got to cut down the last rhododendron on Brownsea Island, marking the end of a 50-year eradication programme to restore the native wooded nature reserve (BBC News, 2011).

Seen from the point of view of climate change, both the present and the future are in a state of dynamic flux. Increasing emissions of greenhouse gases over the last century are now generally accepted as the main drivers of increased global temperature by about 0.5°C since 1970 and changes in the hydrological cycle (IPCC, 2007). Conservative estimates of future climate change indicate global warming of mean surface temperatures by 2–4.5°C, accompanied by changes in rainfall patterns and an increased frequency and severity of extreme environments, such as drought and heat waves (IPCC, 2007). Allen et al. (2010) indicate that at least some of the world's forested ecosystems are already responding to drought- and heat-induced tree mortality, exemplifying the risks of climate change to forests. The speed of temperature change has a global mean of 0.42 km/yr (AIB emission scenario) and ranges from 0.08 km/yr in tropical and subtropical coniferous forests to 1.26 km/yr in flooded grasslands (Loarie et al., 2009). For plants and animals to survive, they must keep pace with and adapt to climates as they move (Pearson, 2006). The rate of northward tree migration during the Holocene after the retreat of the glaciers is estimated at c. 1 km/yr (Pearson, 2006), and may have been as slow as c. 0.1 km/yr if refugia (i.e. areas that remained ice-free during the last glaciation) reseeded the landscape (Loarie et al., 2009). Given the current level of habitat degradation and fragmentation resulting from anthropogenic activities, Loarie et al. (2009) predict that large areas of the globe (28.8 per cent) will require velocities faster than the more optimistic plant migration estimates. In other words, plants will not be able to move fast enough to keep up with changing conditions, resulting in communities being out of step with the local climate, which could lead to widespread decline. Best estimates of species loss indicate extinction of c. 10 per cent of species for each 1°C temperature rise (Fischlin et al., 2007; Convention on Biological Diversity, 2009). Meanwhile some species will persist and others migrate, potentially forming new combinations of species. The future survival of those that remain is further threatened by the release of introduced pests and diseases from climate-limiting factors, such as the occurrence of spring frosts in the UK (e.g. Broadmeadow and Ray, 2005). This suggests that the maintenance of biodiversity requires an increase in the ability of plants and animals to disperse and for flexible and adaptive management plans, which is in some tension with lock-down approaches to biosecurity measures.

This final chapter will function as a conclusion for the edited collection as a whole. We will begin by taking the concept of the future *of* biosecurity, and the future *in* biosecurity practices, broadly conceived, and weave this with themes and threads from the preceding chapters, by way of an overview/review. We will then explore the future of biosecurity through the lens of climate change. This chapter will consider the ways in which climate change challenges biosecurity through the need for species migration; the ways in which climate change increases biosecurity threats; and the use of climate modelling in predicting future invasive patterns. Will climate change demand a new paradigm of ecological management through the growing disparity between 'native' species and suitable ecological conditions? Will we learn to live with and value ecological change or will climate change be used to justify greater biosecurity control, as pest species and diseases escape their barriers and expand their ecological ranges?

Biosecurity and the future

Growing international trade and travel will continue to cause invasions no matter how stringent containment policies are (see Simberloff, Chapter 2), causing the boundary between biosecurity and international trade to be increasingly contested (see Potter, Chapter 8). The very process of making life secure can generate new insecurities such as catastrophic emerging infections (Hinchliffe, Chapter 13). Re-wilding and species reintroductions to restore natural landscapes can create a new suite of (unforeseen) biosecurity problems (Buller, Chapter 12), providing fuel for the nativism debate that underpins many of our conservation practices (see Fall, Chapter 11).

Risk management approaches

The hierarchical risk management process considers all or some of the following: prevention (e.g. border controls), detection, containment (reduction of extent of invasion where eradication is not possible) and restoration. According to Pyšek and Richardson (2010, p. 40) 'preventing the introduction of species with a high risk of becoming invasive is the most cost-effective management strategy'; this would also apply to infectious diseases. This has been promoted by improved accuracy of screening due to advances in databases (e.g. DAISIE, 2009) of introduced species over wide spatial scales and the inclusion of information on invasive capacity. Keller et al. (2007) reported that weed risk assessment screening of ornamental plants could provide Australia with net economic benefits (US$1.67 billion over 50 years) despite the risk of incorrectly rejecting some valuable non-weeds. However, no prevention scheme is completely impenetrable to new arrivals, requiring post-invasion measures to be carried out.

The decision about whether to carry out an intervention requires consideration of ethical questions and the current and future socio-economic impacts of the target invasive species or infectious disease. As far as ethics is concerned, one strand of environmental philosophy holds that individual species have intrinsic value, i.e. that they have value independent of their relationship with, and usefulness for, human beings (Benson, 2000, pp. 85–102). This is the basis of one argument in favour of biodiversity – all species have something like a 'right to life'. The relationship of this argument to biosecurity is equivocal. On the one hand it seems to militate against intervention for two reasons. First, sometimes biosecurity demands the eradication of species in given places and spaces (see the example of rhododendron (*Rhododendron ponticum*) below), and this is in contravention of the 'right to life' principle. The second reason is related to the first, in that the eradication of species is justified in terms of some ideal configuration of species (involving absence as well as presence). Thus the existence of any given species is dependent on its relationship to other species – some 'higher order' arrangement of species in relation to which the presence of any given particular species has to be justified. This robs a particular species of independent value, since its value depends on its relationship with other species. On the other hand, though, the 'right to life' principle could be used to justify intervention where the presence of an invasive species reduces the chances of native species' survival. This is the case for rhododendron, which threatens the survival of the endemic Lundy cabbage (*Coincya wrightii*), and its native endemic invertebrates, found only on Lundy Island in the Bristol Channel (Plantlife, 2010a). Rhododendron also threatens rare lichen and moss communities associated with the internationally important Atlantic hazel and oak woods of Scotland (Plantlife, 2010a).

We can also use invasive non-native rhododendron (*Rhododendron ponticum*) in the UK to illustrate the socio-economic impact of eradication. In 2010, the annual cost of rhododendron to the UK was estimated at *c*. £8.6 million (Williams et al., 2010), and total eradication of rhododendron from woodland in mainland Argyll and Bute and Snowdonia National Park has been estimated to cost £9.6 million and £11 million, respectively (Edwards and Taylor, 2008; Jackson, 2008). There are a variety of management strategies to control rhododendron (Edwards, 2006), with varying degrees of success (Tyler et al., 2006). In the very early stages of rhododendron invasion a small investment in expenditure can prevent the problem getting worse and so save money in the longer term (Edwards and Taylor, 2008). However, the lack of a strategic approach to rhododendron control management can lead to failure of eradication, wasting time and resources. Indeed, 'many eradication efforts [of invasive plants] fail because of poor planning and execution' (Pyšek and Richardson, 2010). Past management programmes were driven by the need to prevent biodiversity loss of native ground flora and ensure successful regeneration of woodlands. However, the seriousness of the situation has been drastically increased by the status of rhododendron as a super-carrier of *Phytophthora ramorum*, the causative agent of Sudden Oak Death.

Edwards and Taylor (2008) used a modified version of Watts et al.'s (2005) least-cost network model (BEETLE) to determine the movement of rhododendrons across the Argyll landscape over 20- and 50-year time periods. This approach has been employed to identify woodlands that provide network routes for focal species to move through landscapes (Watts and Handley, 2010; Watts et al., 2010), and it assumed that rhododendron seeds would disperse furthest through habitats with the least barriers (i.e. open habitat), while vegetative layering would only take place in broadleaf woodlands (Edwards and Taylor, 2008). Over the 50-year time period rhododendron expanded by 55.8 per cent, and occupancy levels in native broadleaf woodlands doubled (Table 14.1). A Woodland Improvement Grants (WIG) calculator (October 2007 release 2.1.1) (Forestry Commission Wales, 2012) was used to estimate the cost of eradicating current and future populations of rhododendron after 20 and 50 years of uncontrolled invasion (Table 14.1). This demonstrates the cost of non-intervention, as delaying eradication for the next 50 years caused the control cost to nearly treble (Table 14.1), exceeding the £25 million set aside by the Department for Environment, Food and Rural Affairs (Defra) in 2009 (FERA, 2009). Jane Kennedy (environment minister) announced that the 5-year programme would 'provide significant funding to help combat these diseases and safeguard our woodlands for the future'.

The financial magnitude of the rhododendron problem goes far beyond available public funds – or at least beyond the funds that governments under late capitalism are generally willing to use to address the problem. Under current

Table 14.1 Current and projected extent of *Rhododendron ponticum* in Argyll and Bute, Scotland, after 0, 20 and 50 years of uncontrolled invasion by vegetative layering and seed dispersal

	Controlled[a]	Uncontrolled rhododendron		
Scenario	0 y (2008)	0 y (2008)	20 y (2028)[b]	50 y (2058)[b]
Area (ha)	704.0	3,950.0	4,851.1	6,248.7
Land base (%)[c]	0.15 (0.8)	0.85 (3.9)	1.1 (5.7)	1.4 (7.7)
Expansion (%)[d]	—	—	22.8	58.2
Control cost (£)	309,515	8,924,664	11,876,744[e]	25,537,955[e]

Source: After Edwards and Taylor, 2008
Notes:
[a] Rhododendron bushes that have undergone control management prior to 2008, may require further treatment to ensure successful eradication
[b] 20- and 50-yr invasion scenarios based on uncontrolled rhododendron source populations only, as past control assumed to be successful
[c] Proportion of land base occupied by rhododendron; values in brackets indicate proportion of native broadleaf woodland occupied by rhododendron
[d] Increase in rhododendron area relative to current extent of uncontrolled rhododendron at year 0
[e] Values assume inflation rate of 3.1%

conditions in which government spending is seen as a necessary evil, Davies and Patenaude (2011) note that the scale of investment required to halt deforestation and biodiversity loss will require contributions from private investors, but this is held back by lack of information on the risks associated with forest propositions worldwide. It is also worth noting that private investors will want to see a return on their investment, and it is not always obvious what this return will be in the context of biodiversity. Even when financial gain and biodiversity are yoked together, as in the developing theory and practice of 'biobanking', there is concern at the implications for nature of viewing it through the lens of capital (Hannis and Sullivan, 2012). The Forest Finance Risk Network (FFRN) established in 2011 provides a platform for 'knowledge exchange between the Natural Environment Research Council-funded and UK research community and end-users in the finance sector on forest-specific risks that affect forest and potential investments' (FFRN, 2012). Rhododendron invasion and Phytophthora are two such risks being assessed.

The fear of an alien future

In an era where change is the only constant, the green and pleasant lands of our past can be viewed with rose-tinted glasses and a time-bound perception of what is virtuous and right (Buller, Chapter 12). Native plants take on saintly properties to be protected from the ravages of advancing aliens, and John Wyndham's 1951 novel *The Day of the Triffids* becomes a possible future reality, with the dominance of the Earth by superweeds – weeds that man can no longer kill (Reed, 2012). Fear of the unknown has been a major driver in biosecurity governance of this unknown future (Rappert and Lentzos, Chapter 10), even though forms of antici-patory regimes to secure public health can create a wave of fear that causes an even greater crisis (Ingram, Chapter 9). Mitigation and restoration of habitats following degradation caused by invasive species is seen as a way of returning ecosystems to their true state; but this can leave legacy effects that increase susceptibility to future invasion (secondary invasions), which can be important contributors to 'invasional meltdown' (Pyšek and Richardson, 2010). Coupled with this is the need to define and select meaningful reference conditions or targets for restoration (e.g. Holmes et al., 2005), which may be out of step with future climates and be a futile and impractical exercise that serves more to appease our sense of guilt than address the incipient needs of the targeted ecosystem.

Biosecurity in a changing climate

Climate change causes a suite of direct and indirect impacts to plants and animals as a result of shifts in the climate itself (temperature, etc.) and associated changes to the disturbance regime that modify and/or regulate the ecosystem. For example, milder winters will result in earlier flush of vegetation that in turn leads to increased survival of herbivores, such as Sika deer (*Cervus nippon*), resulting in damage and increased mortality of commercial forestry trees (Langvatn et al., 1996; Edwards et al., 2008; Table 14.2). The complexity of ecosystem interactions makes it difficult to predict all the possible outcomes and the expected magnitudes of change. Table 14.2 highlights some of the biosecurity threats to UK forests that could be exacerbated by climate change.

Global climate change is predicted to alter the distribution and activity of forest pathogens. In the USA an ecological niche model, CLIMEX, predicts a shift in the distribution of the *P. ramorum* pathogen from 2020 to 2080, likely as a result of changes in heat stress (Venette, 2009). Similarly, Bergot et al. (2004) predict a potential range expansion of *P. cinnamomi* of one to a few hundred kilometres eastwards from the European Atlantic coast within 100 years as a result of increased winter survival of the pathogen due to a predicted 0.5–5°C temperature rise. Biosecurity measures may be insufficient to protect susceptible commercial tree species, such as larch (*Larix* spp.) in the UK. Larch may become another species that the Forestry Commission are unable to use in the future, in much the same way that red band needle blight caused by the fungus Dothistroma needle blight (*Dothistroma septosporum*) sealed the fate of Corsican pine (*Pinus nigra* ssp. *Laricio*) (Forestry Commission, 2007).

Climate change and invasive species represent two of the greatest threats to biodiversity and provisioning of valuable ecosystem services (Burgiel and Muir, 2010). According to Pejchar and Mooney (2010, p. 162) 'ecosystems are life-support systems that provide a suite of goods and services that are vital to human health and livelihood'. Such assets have been categorized by the Millennium Ecosystem Assessment (2005) into four types of ecosystem services: provisioning (e.g. wood products), supporting (e.g. primary production – carbon sequestration),

Table 14.2 Impact of climate change on a selection of biosecurity threats to UK forests

Problem	Status	Forest biosecurity issues	Population response to climate change	Source
Grey squirrels *Sciurus carolinensis*	Invasive pest introduced 1876	Reduced commercial value of damaged trees. Fresh wounds provide entry for pathogens, potentially more susceptible to *Phytophthora*	Population increase due to decreased winter mortality and increased seed availability	Broadmeadow and Ray, 2005; Mayle and Webber, 2012
Sika deer *Cervus nippon*	Invasive pest introduced 1860	Reduced economic value of damaged (browsed, bark stripped, brashed) young trees, prevention of forest regeneration, collapse of Caledonian pinewoods	Population increase due to reduced winter mortality and earlier green-up of spring forage	Langvatn et al., 1996; Edwards et al., 2008
Rhododendron *Rhododendron ponticum*	Invasive plant introduced 1763	Prevents forest regeneration, reduced access to woodlands for operational machinery, super-carrier of *Phytophthora*	Complex. Decreased seedling survival and growth rates due to summer drought, increased colonization rates due to increased disturbance	Edwards and Taylor, 2008; Jackson, 2008
Phytophthora (*Phytophthora. ramorum*, *P. kernoviae*, *P. cinnamomi*)	Invasive soil-borne plant pathogen introduced c.2001	Causative agent of Sudden Oak Death, threatens native trees and commercial larch plantations	Becomes more prevalent and damaging, especially those with higher growth temperature optima (28–30°C), such as *P. cinnamomi*	Broadmeadow and Ray, 2005
Oak decline	Complex tree disorder	Whole-scale decline of native oak woods	Increased incidence due to predicted increase in frequency and severity of summer drought stress	Broadmeadow and Ray, 2005

regulating (e.g. climate regulation) and cultural (e.g. recreational benefits). See Pejchar and Mooney (2010) for an excellent review on the impact of invasive species on ecosystem services. Climate change will impact ecosystem services critical to human societies by altering the balance of invasive species and infectious diseases and threatening long-term food security, public health and wellbeing.

This is especially true the more governments move away from mitigation towards adaptation to climate change. For some time, governments around the world have tried to hold the line at 2°C of warming, based on scientific evidence that any further warming would have runaway effects and then be essentially uncontrollable. However, in the run-up to the Rio+20 conference in Rio de Janeiro in June 2012, Yvo de Boer, former executive secretary of the UN's Framework Convention on Climate Change (UNFCCC), said, 'I think two degrees is out of reach – the two degrees is lost' (de Boer, 2012). This point of view is increasingly accepted, and the signs indeed are that we will exceed what used to be regarded as the limit beyond which runaway climate change would occur. The consequence of this for policy makers is profound, for instead of trying to avoid climate change (mitigation), policy is increasingly directed towards dealing with it as a foregone conclusion (adaptation) (see Davoudi et al., 2009, for example). On the face of it this shift from mitigation to adaptation is not good for biosecurity. This is because climate change prompts the movement and migration of species and pathogens with – as we saw earlier in the chapter – high levels of unpredictability as to outcomes and results. The effects beyond 2°C are of course the subject of discussion and dispute, but a recent US National Research Council report offered a range of predictions as to the effects of warming beyond this point. The report suggests plus or minus 5–10 per cent rainfall variations *per degree*, 5–15 per cent reduction in corn and wheat yields across the world *per degree*, 15–25 per cent reduction in Arctic sea ice *per degree*, and '[A]bout 9 out of 10 summers warmer than the warmest ever experienced during the last decades of the 20th Century over nearly all land areas' (National Research Council, 2011, p. 6). The consequences for biosecurity of these possible changes are very hard to predict, but it seems almost certain that the borders and boundaries that constitute the conditions for the possibility of biosecurity will be ever harder to police. Are we heading towards a future in which we give up on heading off bio-insecurity and opt instead for dealing with its ever-increasingly inevitable consequences?

These consequences can be dramatic. For example, Australia has experienced an increase in food poisoning events and mosquito-borne dengue fever as a result of climate change (Sly, 2011). Issues surrounding social justice and equity will further compound problems experienced in poorer countries, such as the Asia-Pacific region (Sly, 2011). A greater understanding of the ecology of infectious diseases is needed to protect vulnerable populations (Shuman, 2011), which will require better education of the public (Sly, 2011). Furthermore, understanding the interactions of invasive species, disease vectors and pathogens with other drivers of ecosystem change is critical to human health and economic well-being (Crowl et al., 2008). The *Guardian* newspaper reported that not only is pollen getting more potent to hay fever sufferers due to the arrival of 'highly allergenic strains from invasive plants', but 'global warming will cause earlier flowering, possibly extending the hay fever season by six weeks' in the UK (Carrington, 2012). Only by getting a better handle on the complexity of the situation can biosecurity truly address the situation, though the inherent unpredictability of the effects of climate change suggests that policy will be at best a mix of the anticipatory and the reactive, with a likely emphasis on the latter.

Political aspects of climate change and biosecurity

Politics play a vital role in implementing biosecurity, as Part II of this book demonstrates. Perrings et al. (2010, p. 235) note that 'invasive species is a public good' and, 'like all public goods, it will be undersupplied if left to the market', which 'makes it a collective responsibility – a legitimate role of government at many different scales'. In much the same way, in order to respond to future climates, changes to industry practices and government policy may be required (Cooperative Research Centre 2012). Outhwaite (Chapter 5) highlights the problems of developing biosecurity legislation for national and international legal regimes, which requires willingness for countries to engage. The climate change denial movement, headlined by the George W. Bush administration, which claimed that 'CO_2 is not a pollutant' and attempted to downplay the scientific evidence for climate change in US documents (Revkin, 2005), demonstrates the sometimes problematic role of politics.

Invasive species are a global concern with local consequences that require policy and legislation at the national and international level. The main problem is that 'biosecurity policies and strategies are being implemented without adequate conceptualisation and verification of keystone assumptions' (Pyšek and Richardson, 2010, p. 48) and fail to take into account consequences of climate change.

The scale of responsibility is determined in part by the type of pathway – national regulations are needed for release and escape of invasive species, whereas international policies address contaminant, stowaway and dispersal pathways (Hulme et al., 2008). In the UK, numerous government-led initiatives have been put in place to ensure biosecurity of forestry (e.g. Defra and Forestry Commission, 2011; Forestry Commission, 2012). Closure of the Global Invasive Species Program (GISP) Secretariat sends a rather mixed message as to the importance of invasive species; although the website has been relaunched there are no funds to produce new GISP publications or update the website on a regular basis (GISP, 2012).

Role of technology and advancements in science

For biosecurity measures to be timely and cost-effective, up-to-date information needs to be available to monitor the movement and progress of pests and infectious diseases. In 2009 'internet chatter' in the form of blogs and twitter stories helped to track the swine flu outbreak (Madrigal, 2009). Dugas et al. (2012) believe monitoring internet search traffic about influenza may better prepare hospital emergency rooms to a surge in sick patients than outdated government flu case reports.

Similar advances are being made in the field of monitoring. Traditional manual vegetation surveys to detect invasive plant species require identification of species on the ground by phenotype (physical appearance) (Schmidt and Skidmore, 2003), which is significantly prone to human error as many species are similar even at close range to the untrained eye. For example, cherry laurel (*Prunus laurocerasus*) is visually similar to invasive rhododendron (*R. ponticum*) and, according to the Landscape Ecology team at Forest Research, 'can be confused with the latter in long-distance sight mapping' (Amy Eyecott, pers. comm., 2011). Better, more user-friendly identification guides for plants and seeds, along with new high-tech diagnostic tools for microorganisms, such as gene probes, DNA barcoding and acoustic sensors (Pyšek and Richardson, 2010), are dramatically improving the detection phase of risk management. Extensive research is being carried out to quantify the probability that a given survey technique will detect a target species if it is present (Hayes et al., 2005). For example, a radiospectrometry study by Taylor et al. (2013) in standardized dark room conditions generated a 94 per cent success rate in discerning rhododendron from cherry laurel.

The new age of remote sensing equipment provides a reliable, rapid and comprehensive alternative technique, which is increasingly taking a leading role in detection and monitoring. A Caerphilly County Borough Council-led project, involving five local authorities and the Environment Agency Wales, used LiDAR (Light Detection and Ranging), aerial photography and an object-oriented imaging software, eCognition (Trimble Navigations Ltd), to identify the likely locations of Japanese knotweed (Jarman, 2010). This is one of numerous feasibility studies carried out to assess the ability of remote sensing to detect and monitor invasive and nuisance species (e.g. Underwood et al., 2003; Waldrop, 2010; Taylor et al., 2013).

Digital spatial mapping in geographic information systems is also enabling the development of risk maps that take into account the effect of fragmented landscapes and evaluation of critical uncertainty thresholds of invasion risk (Koch et al., 2009). For example, Edwards and Taylor (2008) produced an integrated model of both rhododendron vegetative spread and airborne seed dispersal that took into account additional dispersal agents such as river, road and rail networks in order to determine overall risk of rhododendron invasion.

Paradigm shift – nativism to novel ecosystems

The escalating scale of bioinvasions, species synergies and interactions with global change factors thwart restoration efforts and challenge the ideals of conserving native species out of phase with future climates. There is resistance to change, as demonstrated by the Institute of Ecology and Environmental Management spring 2011 conference on 'Invasive Species: New Natives in a Changing Climate', which maintained the ethos 'native good – invasive bad' despite the supposed remit of the conference. However, the nativism paradigm is undergoing a period of change, and key to this is the promotion of the integrity of the ecosystem rather than focusing on individual native species that currently occupy a particular ecological niche. Such 'novel ecosystems'… are 'comprised of species that occur in combinations and relative abundances that have not occurred previously at a given location or biome' (Pyšek and Richardson, 2010, p. 46). A classic example is the change in forest ecosystem assemblages at the southern foot of the Alps in response to climate change. Historical deciduous broad-leaved forest has been superseded by evergreen broad-leaved forest, including an introduced tropical hemp palm (*Trachycarpus fortune*) (Walther et al., 2007). Perhaps attempts to control these changes are like 'just spitting into the wind of invasive species blowing across our restoration sites' (Allison, 2011, p. 265). It's time for a new perspective, to quote Thomas and Ohlemüller (2010, p. 26): 'we can no longer presume that the arrival of species from other regions and countries should be regarded as negative'. The influx of Mediterranean and tropical plant species that are filling ecological niches vacated by native plants unable to cope with changing climates have the potential to increase the resilience of the ecosystem to change and prevent widespread complexes of communities.

Conclusion

We believe climate change will demand a new paradigm of ecological management through the growing disparity between 'native' species and suitable ecological conditions, and there are already indications of this being embraced, albeit slowly. While the threat of climate change may be a catalyst for greater biosecurity control (e.g. pre-border, etc.) this is not going to prevent an influx of pests and diseases – absolute control of the effects of anthropogenic climate change is impossible. Financially, eradication programmes are not a cost-effective means of control once an introduced population is established, and the legacy of restoration may encourage future unwanted arrivals. The future of our ecosystems may well depend on these newly arrived species, as they have the potential to increase resilience and maintain ecosystem services in the long run.

References

Albrecht, G. (2005) 'Solastalgia, a new concept in human health and identity', Philosophy Activism Nature, vol. 3, pp. 41–44.

Allen, C. D., Macalady, A. K., Chenchouni, H., Bachelet, D., McDowell, N. et al. (2010) 'A global overview of drought and heat-induced tree mortality reveals emerging climate change risks for forests', *Forest Ecology and Management*, vol. 259, pp. 660–684.

Allison, S. K. (2011) 'The paradox of invasive species in ecological restoration: do restorationists worry about them too much or too little?' in I. D. Rotherham and R. A. Lambert (eds) *Invasive and introduced plants and animals: human perceptions, attitudes and approaches to management*, Earthscan, London, pp. 265–275.

BBC News (2011) 'Brownsea Island holds raffle to cut last rhododendron', www.bbc.co.uk/news/uk-england-dorset-13871992, accessed 7 September 2012.

Benson, J. (2000) *Environmental ethics: an introduction with readings*, Routledge, London and New York.

Bergot, M., Cloppet, E., Pérarnaud, V., Déqué, M., Marçais, B. and Desprez-Loustau, M-L. (2004) 'Simulation of potential range expansion of oak disease caused by Phytophthora cinnamomi under climate change', *Global Change Biology*, vol. 10, no. 9, pp. 1539–1552.

Broadmeadow, M. and Ray, D. (2005) *Climate change and British woodland*, Forestry Commission Information Note 069, www.forestry.gov.uk.

Burgiel, S. W. and Muir, A. A. (2010) 'Invasive species, climate change and ecosystem-based adaptation: addressing multiple drivers of global change', Global Invasive Species Programme (GISP), Washington, DC, and Nairobi, Kenya.

Carrington, D. (2012) 'Climate change will extend hay fever season by six weeks, report warns', *Guardian Online*, 11 September, www.guardian.co.uk/environment/2012/sep/11/climate-change-hayfever, accessed 11 September 2012.

Convention on Biological Diversity (2009) *Draft findings of the ad hoc technical expert group on biodiversity and climate change*, Convention on Biological Diversity, Montreal.

Cooperative Research Centre (CRC) for National Plant Biosecurity (2012) 'CRC10071: climate change', www.crcplantbiosecurity.com.au/project/crc10071-climate-change, accessed 31 July 2012.

Crowl, T. A., Crist, T. O., Parmenter, R. R., Belovsky, G. and Lugo, A. E. (2008) 'The spread of invasive species and infectious diseases as drivers of ecosystem change', *Frontiers in Ecology and Environment*, vol. 6, no. 5, pp. 238–246.

DAISIE (2009) *Handbook of alien species in Europe*, Springer, Berlin.

Davies, S. and Patenaude, G. (2011) 'Addressing the forest science versus investment nexus: can a more holistic understanding of risks bridge the gap?', *Carbon Management*, vol. 2, no. 6, pp. 613–616.

Davoudi, S., Crawford, J. and Mehmood, A. (2009) *Planning for climate change: strategies for mitigation and adaptation for spatial planners*, Earthscan, London.

De Boer, Y. (2012) 'Two degree target "out of reach": former UN climate chief', www.timeslive.co.za/scitech/2012/03/27/two-degree-target-out-of-reach-former-un-climate-chief, accessed 4 September 2012.

Defra and Forestry Commission (2011) 'Action plan for tree health and plant biosecurity', report available for download from www.defra.gov.uk/food-farm/crops/plant-health/, accessed 20 April 2012.

Dugas, A. F., Hsieh, Y-H., Levin, S. R., Pines, J. M., Mareiniss, D. P. et al. (2012) 'Google flu trends: correlation with emergency department influenza rates and crowding metrics', *Clinical Infectious Diseases*, vol. 54, no. 4, pp. 463–469.

Edwards, C. (2006) 'Managing and controlling invasive rhododendron', *Forestry Commission Practice Guide 017*, Forestry Commission, Edinburgh, 36pp.

Edwards, C. and Taylor, S. L. (2008) 'A survey and strategies appraisal of rhododendron invasion and control in woodland areas in Argyll and Bute', a contract report for Perth Conservancy, Forestry Commission, prepared by Forest Research, executive summary available at www.forestry.gov.uk/pdf/Argyll_Bute_rhododendron_report_2008_exec. pdf.pdf/$FILE/Argyll_Bute_rhododendron_report_2008_exec.pdf.pdf, accessed 11 September 2012.

Edwards, C., Taylor, S. L. and Peace, A. (2008) 'Determining the risk of pinewood deterioration based on tree size (DBH) structure and regeneration density data', a contract report for the Deep Commission, Scotland, prepared by Forest Research, March 2008, available at www.snh.gov.uk/publications-data-and-research/publications/search-the-catalogue/publication-detail/?id=1707, accessed 11 September 2012.

FERA (Food and Environment Research Agency) (2009) 'Fera to manage £25 million Defra campaign to fight woodland disease', www.fera.defra.gov.uk/showNews.cfm? id=393, accessed 21 February 2013.

FFRN (2012) 'Forest finance risk network', http://xweb.geos.ed.ac.uk/~gpatenau/ ForestFinance/Forest_Finance_Risk_Network/Homepage.html, accessed 18 July 2012.

Fischlin, A., Midgley, G. F., Price, J. T., Leemans, R., Gopal, B. et al. (2007) 'Ecosystems, their properties, goods and services', in M. Parry, O. F. Canziani, J. P. Palutikof, P. J. van der Linden and C. E. Hanson (eds) *Climate change 2007: Impacts, adaptation and vulnerability*, contribution of Working Group II to the Fourth Assessment, Report of the Intergovernmental Panel on Climate change, Cambridge University Press, Cambridge.

Forestry Commission (2007) 'Forestry Commission suspends Corsican pine planting', News Release 9339, www.forestry.gov.uk/newsrele.nsf/AllByUNID/5E8A8246143878B3802 5729500510FB6, accessed 1 August 2012.

—— (2012) 'Forestry Commission biosecurity programme board', www.forestry.gov.uk/ forestry/INFD-7XUGL2, accessed 16 April 2012.

Forestry Commission Wales (2012) 'Calculating woodland improvement grants', www. forestry.gov.uk/forestry/INFD-6LCJM4, accessed 27 April 2012.

GISP (2012) 'Global invasive species programme', www.gisp.org/, accessed 2 August 2012.

Gray, R. (2010) 'Mortgages refused over invasive weed', *Daily Telegraph* [online] 13 March 2010, www.telegraph.co.uk/property/propertynews/7436431/Mortgages-refusedover-invasive-weed.html, accessed 6 March 2012.

Hannis, M. and Sullivan, S. (2012) *Offsetting nature? Habitat banking and biodiversity offsets in the English land use planning system*, www.greenhousethinktank.org/files/greenhouse/ home/Offsetting_nature_inner_final.pdf, accessed 4 September 2012.

Hayes, K. R., Cannon, R., Neil, K. and Inglis, G. (2005) 'Sensitivity and cost considerations for the detection and eradication of marine pests in ports', *Marine Pollution Bulletin*, vol. 50, pp. 823–834.

Holmes, P. M., Richardson, D. M., Esler, K. J., Witkowski, E. T. F. and Fourie, S. (2005) 'A decision-making framework for restoring riparian zones degraded by invasive alien plants in South Africa', *South African Journal of Science*, vol. 101, pp. 553–564.

Hulme, P. E., Bacher, S., Kenis, M., Klotz, S., Kühn, I. et al. (2008) 'Grasping at the routes of biological invasions: a framework for integrating pathways into policy', *Journal of Applied Ecology*, vol. 45, pp. 403–414.

IPCC (2007) 'Climate change 2007: the physical science basis', in S. Solomon, D. Qin, M. Manning, Z. Chen, M. Marquis, K. B. Avery, M. Tignor and H. L. Miller (eds) *Contribution of the Working Group I to the Fourth Assessment, Report of the*

Intergovernmental Panel on Climate Change, Cambridge University Press, Cambridge, 996pp.

Jackson, P. (2008) 'Rhododendron in Snowdonia and a strategy for its control', Snowdonia National Park Authority, Penrhyndeudraeth, Gwynedd, UK, www.eryri-npa.gov.uk.

Jarman, M. (2010) 'Identification of invasive weed species through the use of remote sensing techniques', Environment Systems Ltd, in Association for Geographic Information Cymru Annual Conference, 1 December 2010, Cardiff City Hall, Wales, www.agi.org.uk/storage/events/20101201-Cymru/MarkJarman1.pdf, accessed 11 November 2011.

Keller, R., Lodge, D. M. and Finnoff, D. (2007) 'Risk assessment for invasive species produces net bioeconomic benefits', *Proceedings of the National Academy of Sciences (PNAS) of the United States*, vol. 104, pp. 203–207.

Koch, F. H., Yemshanov, D., Mckenney, D. W. and Smith, W. D. (2009) 'Evaluating critical uncertainty thresholds in a spatial model of forest pest invasion risk', *Risk Analysis*, vol. 29, pp. 1227–1241.

Langvatn, R., Albon, S. D., Burkey, T. and Cluttonbrock, T. H. (1996) 'Climate, plant phenology and variation in age of first reproduction in a temperate herbivore', *Journal of Animal Ecology*, vol. 65, pp. 653–670.

Loarie, S. R., Duffy, P. B., Hamilton, H., Asner, G. P., Field, C. B. and Ackerly, D. D. (2009) 'The velocity of climate change', *Nature*, vol. 462, pp. 1052–1055.

Madrigal, A. (2009) 'Tracking internet chatter helps spot swine flu outbreak', *Wired Science*, 27 April 2009, www.wired.com/wiredscience/2009/04/swinefluchatter/, accessed 1 August 2012.

Mayle, B. and Webber, J. (2012) 'Are grey squirrels implicated in spreading *P. ramorum?*', www.forestry.gov.uk/forestry/infd-8rcmkv, accessed 27 April 2012.

Millennium Ecosystem Assessment (MEA) (2005) *Ecosystems and human well-being synthesis*, Island Press, Washington, DC.

National Research Council (2011) *Warming world: impacts by degree*, http://dels.nas.edu/resources/static-assets/materials-based-on-reports/booklets/warming_world_final.pdf, accessed 4 September 2012.

Pearson, R. G. (2006) 'Climate change and the migration capacity of species', *Trends in Ecology and Evolution*, vol. 21, pp. 111–113.

Pejchar, L. and Mooney, H. (2010) 'The impact of invasive alien species on ecosystem services and human well-being', in C. Perrings, H. Mooney and M. Williamson (eds) *Bioinvasions and globalization: ecology, economics, management, and policy*, Oxford University Press, Oxford, pp. 161–182.

Perrings, C., Burgiel, S., Lonsdale, M., Mooney, H. and Williamson, M. (2010) 'Globalisation and bioinvasions: the international policy problem', in C. Perrings, H. Mooney and M. Williamson (eds) *Bioinvasions and globalization: ecology, economics, management, and policy*, Oxford University Press, Oxford, pp. 235–250.

Plantlife (2010a) 'Rhododendron (*Rhododendron ponticum*)', www.plantlife.org.uk/wild_plants/plant_species/rhododendron, accessed 7 September 2012.

—— (2010b) *Here today, here tomorrow? Horizon scanning for invasive non-native plants*, 19pp, www.plantlife.org.uk/uploads/documents/Here_today_here_tomorrow_2010_summary.pdf.

Pyšek, P. and Richardson, D. M. (2010) 'Invasive species, environmental climate and management, and health', *Annual Review of Environment and Resources*, vol. 35, pp. 25–55.

Reed, G. (2012) 'A growing problem: notes from the "superweed" summit', http://grist.org/industrial-agriculture/a-growing-problem-notes-from-the-superweed-summit/, accessed 31 July 2012.

Revkin, A. C. (2005) 'Bush aide edited climate reports', *New York Times*, 8 June 2005, www.nytimes.com/2005/06/08/politics/08climate.html?_r=1, accessed 31 July 2012.

Robinet, C. and Roques, A. (2010) 'Direct impacts of recent climate warming on insect populations', *Integrative Zoology*, vol. 5, pp. 132–142.

Roques, A. (2010) 'Alien forest insects in a warmer world and a globalised economy: impacts of changes in trade, tourism and climate on forest biosecurity', *New Zealand Journal of Forest Science*, vol. 40 (suppl.), pp. S77–S94.

Schmidt, K. and Skidmore, A. (2003) 'Spectral discrimination of vegetation types in a coastal wetland', *Remote Sensing of Environment*, vol. 85, pp. 92–108.

Shuman, E. K. (2011) 'Global climate change and infectious diseases', *International Journal of Occupational and Environmental Medicine*, vol. 2, no. 1, pp. 11–19.

Sly, P. D. (2011) 'Health impacts of climate change and biosecurity in the Asian Pacific region', *Reviews on Environmental Health*, vol. 26, no. 1, pp. 7–12.

Taylor, S. L., Hill, R. and Edwards, C. (2013) 'Characterising invasive non-native *Rhododendron ponticum* spectra signatures with radiospectrometry in the laboratory and field: potential for remote mapping', *ISPRS Journal of Photogrammetry and Remote Sensing*, vol. 81, pp. 70–81.

Thomas, C. D. and Ohlemüller, R. (2010) 'Climate change and species' distributions: an alien future?' in C. Perrings, H. Mooney and M. Williamson (eds) *Bioinvasions and globalization: ecology, economics, management, and policy*, Oxford University Press, Oxford, pp. 19–29.

Tyler, C., Pullin, A. S. and Stewart, G. B. (2006) 'Effectiveness of management interventions to control invasion by *Rhododendron ponticum*', *Environmental Management*, vol. 37, pp. 513–522.

Underwood, E., Ustin, S. L. and Dipietro, D. (2003) 'Mapping non-native plants using hyperspectral imagery', *Remote Sensing and Environment*, vol. 86, pp. 150–161.

Venette, R. C. (2009) 'Implication of global climate change on the distribution and activity of *Phytophthora ramorum*', *2009 USDA Research Forum on Invasive Species*, pp. 58–59.

Waldrop, T. (2010) 'Naked eyes and hyperspectral images build fuel maps in the southern Appalachian mountains', *Fire Science Brief*, vol. 117, pp. 1–6.

Walther, G. R., Gritti, E. S., Berger, S., Hickler, T., Tang, Z. Y. and Sykes, M. T. (2007) 'Palms tracking climate change', *Global Ecology and Biogeography*, vol. 16, pp. 801–809.

Watts, K. and Handley, P. (2010) 'Developing a functional connectivity indicator to detect change in fragmented landscapes', *Ecological Indicators*, vol. 10, pp. 552–557.

Watts, K., Eycott, A. E., Handley, P., Ray, D., Humphrey, J. W. and Quine, C. P. (2010) 'Targeting and evaluating biodiversity conservation action within fragmented landscapes: an approach based on generic focal species and least-cost networks', *Landscape Ecology*, vol. 25, pp. 1305–1318.

Watts, K., Humphrey, J. W., Griffiths, M., Quine, C. and Ray, D. (2005) 'Evaluating biodiversity in fragmented landscapes: principles', Forestry Commission Information Note 073, www.forestry.gov.uk.

Williams, F., Eschen, R., Harris, A., Djeddour, D., Pratt, C. et al. (2010) *The economic cost of invasive non-native species on Great Britain*', CABI, Wallingford, UK.

INDEX